The Spirit of Science

The Spirit of Science

From Experiment to Experience

Edited by David Lorimer

CONTINUUM • NEW YORK

1999

The Continuum Publishing Company
370 Lexington Avenue
New York, NY 10017

Printed in the United States of America

First published in the United Kingdom by Floris Books, Edinburgh

Library of Congress Cataloging-in-Publication Data

The spirit of science : from experiment to experience / edited by David Lorimer.
    p. cm.
    Includes index.
    ISBN 0-8264-1174-6 (pbk.)
    1. Science—Philosophy. 2. Religion and science. 3. Gaia hypothesis.
4. Mysticism. I. Lorimer, David, 1952–   .
Q175.S677   1999
501—dc21   98-32282
                                               CIP

Contents

Consciousness and Psychology

Mysticism and Spirituality

Acknowledgments

I would like to thank all the people who helped in the production of this book: the authors, who assiduously corrected and amended the transcripts; Jeanette Griffiths, who typed up the transcripts; Christopher Moore for his editorial assistance; and, most of all, Kevin Ashbridge for the long hours put into editing a number of the scientific lectures.

Every effort has been made to contact copyright holders of quoted material, and any lack of acknowledgment will be corrected where possible in subsequent editions.

D.L.

Preface

MALCOLM LAZARUS

A suggestion made over twenty years ago that scientists and mystics might have something in common would have been considered strange. Scientists were thought to be concerned with verifiable facts and designed experiments which when duplicated would have similar results and could be described in precise language. The language, however, that sub-nuclear physicists were using to describe the behaviour of particles of matter in an atomic accelerator was beginning to look exceedingly misty around the edges. Particles had a tendency to exist. They did not always behave like discrete bits of matter — often like waves.

Like mystical experience they were unpredictable and could not be placed in time or space. The where and when of them was elusive. Both Fritjof Capra and Lawrence Le Shan had picked this up and written about it in an immensely intriguing way. This seemed to fit in with the education programme of Sir George Trevelyan's fledgling Wrekin Trust for which I had taken increasing responsibility since joining him in 1974. The Wrekin Trust was concerned with the 'Spiritual Nature of Man and the Universe' and suggested new ways for people to look at themselves and their society, with the possibility of formulating a new consciousness uniting the material and the spiritual. This was a concept out of which a new purpose and aim for individual lives might emerge and indeed a new society might be born. An impulse no less strong today that it was twenty years ago. So exploring the relationship between mystics and scientists seemed like a fruitful path to follow, full of possibilities.

George Trevelyan was fond of saying that political change only scraped the surface and that the only change of real value was an inner willed change in consciousness. I saw, though not all at once, that a new rationale for experiencing ourselves through the exploration of mystical, scientific and artistic experience could contribute to such a change.

For myself, the Mystics and Scientists conferences, starting in 1978, became increasingly intellectually challenging as my own mind development came as a result of mystical and psychospiritual experience rather than the study of science. Stretching my mind to embrace concepts that were being articulated by some of the great minds of our time was a spiritual discipline in itself, and presenting people with a compelling reason to travel, often thousands of miles, to participate, stretched it even further. I often felt my mind splitting open like that of a student in a Zen monastery whose teacher has asked the unanswerable question that leads to illumination.

I'll recount just a few unusual events, special meetings and interesting ideas to give you a flavour of the atmosphere around these great conferences. They are not comprehensive and omit recollections of many of the great people who have enriched my life and delighted the delegates.

At MS2, when chairing the question and answer session, I had Professor Hasted, who was doing research into metal bending and teleportation on one side, and Guru Raj Ananda Yogi on the other — the Mystic leaned across and from the ends of his fingers produced tiny lumps of what looked like hard black clinker which he handed to Hasted who folded it up carefully in white paper whispering it was Vibouti and that he would analyse it. Months later I asked what the analysis had shown. 'When I opened the paper,' he said 'The Vibouti had vanished!!'

It was during that conference that I met the Tibetan incarnate Lama Sogyal Rinpoche who reported on it to His Holiness the Dalai Lama, who invited me to a private audience in Switzerland. At that time he was physically in his prime and radiated a very strong energy field. It was like sitting next to a warm stove. He seemed totally aware. His concentration was entirely focused. Amongst other matters we discussed problems I had heard about and encountered amongst Westerners who had spent long period meditating in Tibetan monasteries on energies depicted in Tibetan iconography — feeding images alien to their tradition into the unconscious. Instead of resonating and integrating — meshing if you like — with energies and images residing there, this produced disharmony and conflict.

For Tibetans such practices would have led to deep experiences of harmony and a reconnection with the spiritual line. In Westerners there was a danger that the images arising out of the unconscious were so different as to produce serious emotional and mental dislocation. His Holiness recognized the problem and said he would try to do something about it. The monks at that time clearly could not cope. Then, with a twinkle in his eye he said, 'Malcolm, you should understand that even Tibetans go mad!' — a phrase that I never forgot and that had a valuable effect on my own teaching. As a result of that meeting I also began to understand why intermarriage between people of different spiritual bloodlines was often proscribed. It has also been my experience that blood, memory and DNA carry their own racial resonances to the sacred.

Around MS4 we reached a high point with Karl Pribram and Charles Tart coming over from the States to join Glen Schaefer, The Ven Sumedho and Warren Kenton. I remember Arthur Ellison saying that this would be a hard act to follow. Amongst a panel of brilliant speakers it was Glen, Professor of Ecological Physics at Cranfield, who received a standing ovation after his talk on the necessity for a holistic philosophy of nature and the failure of chemical warfare on insects. He outlined several inconsistencies in evolutionary theory and spoke of the delicate balance on which the created world depends and how Man in his efforts to control pests through mass killing by insecticides was trifling with material forces that he could not control. These forces would resist in order to survive and maintain the ecological network. Man's purpose on earth, he argued, was redemptive in nature, which meant working with insects instead of destroying them. This was in any event impossible because they had always evolved faster than our abilities to counteract them. He seemed like a prophet crying passionately in the wilderness that humankind would cause pests to evolve in his own image. We would literally create our own pestilence. Glen influenced the conferences in other ways as well, stressing the importance of providing a proper honorarium for lecturers to make it worth their while to spend time reworking or developing new material for the conferences, and helping them to become more prestigious occasions. He also acted as informal scientific adviser and was insistent that speakers were chosen with a high record of academic achievement. It was through his influence that Sir John Eccles, the great neurophysiologist and Nobel Laureate, agreed to participate in the series.

Eccles argued that mind had an existence independent of the brain; that the uniqueness of individual personalities did not depend on genetics, and that science had gone too far in breaking down Man's belief in his spiritual greatness. He considered that we were creatures with some supernatural meaning and endowed with purpose.

Some time later, my wife and I were having a meal at a restaurant in the Austrian Tyrol with Arnold Keyserling from Vienna. We were talking about ideas for developing a spiritual curriculum derived from the medieval concept of the University which was concerned to find methods and systems of knowledge leading to wholeness or union with the One — as the term *Universus* ('turned to the one') implies. Professors were people who professed to be able to help students to do just that: a far cry from today's career orientated curriculum with the Liberal Arts being treated as academic subjects. Such arts were originally concerned with personal development — the Arts of Liberation. This conversation was also the inspiration for MS8: Music, Mathematics and Consciousness.

Well, Sir John Eccles was dining in the same restaurant with an elderly, very small, almost gnomelike man — and that's when I met the legendary Karl Popper with whom Eccles had recently developed the theory of dualist interactionism, saying that 'The theory offered valuable insights into the higher levels of human experience that cannot be accommodated to materialistic theories of mind.'

Eccles led to David Bohm talking at MS6. He was head of the Department of Theoretical Physics at the University of London and was recognized as one of the world's leading researchers into quantum theory. He was convinced that matter and consciousness formed an unbroken whole. In the introduction to his book *Wholeness and the Implicate Order,* he says that the widespread and pervasive distinctions between people — race — nation — and so on, now preventing humankind from working together for the common good and indeed for survival originate in a kind of reductionist thought that treats things as being inherently divided — disconnected, broken up into smaller constituent parts.

During the same conference Keyserling also drew our attention to the way in which language fragments reality. The whole *subject–verb–object* structure of sentences implies that action arises in a separate entity, an object that is then described as if it were then static. Bohm cites a good example, 'We say *it* is raining ... Where is the *it?*' 'This structure of language affecting as it does the way we think tends to divide things into separate entities which are thought of as being fixed and static. It is therefore not surprising that we have a scientific world view in which everything is regarded as being ultimately constituted out of a set of basic particles of a fixed nature.' Once we arrive at a philosophy of wholeness, we are going to need a language to describe it. Otherwise we shall continue with the sort of schizophrenic situation in which we see objects as separate, think about them as being separate with the ordinary mind, yet know that this is not the case.

I want to return to Glen Schaefer who used to talk passionately about the impossibility of dealing constructively with pests until we had come to terms with our own inner greed and fear and the nature of our own inner pestilence. For most scientists inner exploration and self-consciousness are not yet part of the training. This omission is beginning to take on the dimensions of a tragedy. Witness our misplaced adversarial confrontation with infections and the resultant misuse of antibiotics. Yet science has known for several decades that the consciousness of the experimenter affects the experiment. Hence the importance of scientists and mystics coming together to discuss similarities in their twin approaches to truth and knowledge. It is even more important to take steps to transform that knowledge into experience. Today scientists with only a modicum of self-knowledge are working with vast creative as well as potentially destructive powers in many fields. Mystics can help them gain the wisdom that is so necessary.

At the twentieth conference I spoke about what I had learned through teaching a course for twenty years entitled 'The Transformational Journey and the Mystical Experience.' This has convinced me that all of us are pre-programmed biologically to experience mystical and ecstatic states and that everyone with the motivation to have these experiences can have them, with help.

It is, however, vital to do the groundwork and spend time loosening the bonds of attachment to the 'little ego' — the person we think we are. We are aware of sub-personalities in Transpersonal Psychology and the domination of the conscious mind. Additionally, it is necessary to come to terms with the shadow and learn to accept and even embrace the content of our individual dustbins — or perhaps 'cesspits' is a more creative analogy. If these two stages are neglected, people run the risk of opening the inner gates in an uncontrolled way and being overwhelmed by a torrent of material that could cripple them.

Spiritual technologies exist to catalyse the transforming process of mystical experiences. It is actually quite an easy thing to do. So the question now becomes: how can we focus our consciousness in such a way that when we enter those states, something of value occurs? There are, of course, many ways to do this and different spiritual traditions do it in different ways.

In 1995 I took a group of people to Turkey, all of whom had had mystical experience. Sufi Sheikhs, with great courage after testing us quite severely, let us witness their practices of Zihkr. Their whole focus of a life led in love and service of the Beloved was their preparation. So when, with a combination of breathing, chanting and swaying, they entered the mystical state — or at the least an altered state of consciousness — they could open themselves or be opened to the ineffable. The results were apparent in the beauty of their being and the sublime energy that flooded the room.

Mystical states sometimes seem to be biological. The fountain of eternal life, for example, with everything it may bring seems to have something to do with the action of the heart and the pumping of blood. Experience of recollection of evolutionary states and the emotions that accompany them seem to be about entering into the DNA, which carry memories of the evolution of our race.

A new and vital spirituality might well emerge as a development of spiritual technologies taken out of the context of the religious traditions that appear to have lost the ability to use them. It is my hope that scientists may be drawn in the future to courses, run by mature organizations grounded in reality like the Scientific and Medical Network, on 'Exploring Spiritual Technologies.' It has a nice ring to it. They would, of course, be experiential.

Finally, I want to say that although I am credited with making these conferences happen, it is really the delegates and speakers whose support makes it all possible. That support took me on a fabulous journey for what became the best years of my life which are still continuing. It gave me access to some of the most interesting people in the world. Their willingness to share that journey with me and to harvest some of its fruits along the way gave me the opportunity to explore the greatest questions of all: the nature of life and my own humanity. I thank them all. I am deeply grateful.

David Lorimer has now successfully brought together twenty years of work in producing this major anthology which challenges us to dwell on thoughts not often expressed so cogently. The reader will be richly rewarded.

Introduction:
From Experiment to Experience

DAVID LORIMER

David Lorimer has been Director of the Scientific and Medical Network since 1986. He was educated at Eton and the Universities of St Andrews and Cambridge. After a spell in merchant banking, he spent a number of years teaching modern languages and philosophy at Winchester College before assuming his present position. He is author of Survival? Body, Mind and Death in the Light of Psychic Experience *(1984) and* Whole in One *(1990). He is also editor of* The Circle of Sacred Dance: Peter Deunov's Paneurythmy *(1991),* Prophet for our Times *(1991) and* Gems of Love *(1994). The second part of this introduction is based on a lecture given at the 1997 Mystics and Scientists conference.*

A brief overview

I attended my first Mystics and Scientists conference in 1981, at a time when I was teaching down the hill at Winchester College. It was a heady, even electrifying experience. There was a great sense of excitement that two important fields of human endeavour were being considered together by mystics and scientists who recognized that neither had the full picture. I remember in particular a remark by Charles Tart to the effect that the conflict between mysticism and science was a conflict between dogmatists who were second-rate mystics and second-rate scientists. Those in the front rank recognized the complementary approaches to the same reality. When one considers some of the great physicists of twentieth century science like Einstein,

Heisenberg, Pauli, Bohr and Eddington, one can appreciate the force of his remark.

The very first Mystics and Scientists conference was held in 1978, just two years after the publication of Fritjof Capra's *The Tao of Physics* and the year before Gary Zukav's *The Dancing Wu Li Masters*. The idea of a confluence between mysticism and science was just being tentatively put forward and many were struck by parallel language being used to speak about the underlying nature of reality. I remember attending a lecture in Cambridge by Fritjof Capra in that same year, where the mood was one of openness to new horizons. The fundamental issue can be expressed in the question: what is the relationship between the one and the many? A supplementary question asks if the one of the scientist is the same as the one of the mystic.

Both mystics and scientists posit a unity underlying the multiplicity of the world, the ground of reality and being or consciousness. This quest for unity or unification is well expressed in the title of Renée Weber's classic 1986 book *Dialogues with Scientists and Sages: the Search for Unity*. Readers will also find this fully discussed in Kurt Dressler's piece below. By 1983, Fritjof Capra and Ken Wilber were engaged in an exchange on this very issue. Capra argued that the unity in question was ultimately identical, while Wilber maintained that the unity underlying the physical world was one of subscendence, while unitive consciousness implied a transcendence. It may well be true that these are ultimately identical, but I find Wilber's distinction a useful one to bear in mind when teasing out the meaning of unity advanced in these different contexts.

Mystics and Scientists twenty years on

A quick glance at the hundred or so lecturers at the past twenty conferences reveals that they are at the forefront of new thinking in their respective fields and are the authors of many of the most widely influential books. Many readers of this book will surely find that they own at least a shelf of books by Mystics and Scientists lecturers! I calculate that just over fifty of the lecturers have produced a total of over 180 books, most of which are in my own library. I believe that the conferences themselves have contributed towards the changes of thinking that we have witnessed, but that this has been amplified considerably by the individual contributions of the lecturers. With an average attendance of nearly three hundred people a year, many of whom have attended previous conferences, the total attendance must be well over 3,000 people. One then needs to add those who did not attend but who subsequently ordered tapes or videos (see Editor's Note on p.357). Finally, if one factors in the millions of people who have read some of the 180 key books, the outreach of the conference is indirectly very considerable.

What are the key developments in the fields covered by Mystics and Scientists conferences over the past twenty years?

In the field of mysticism and spirituality, there has been a growing interest in Asian religions, comparative religion, meditation and personal spiritual development. Dom Bede Griffiths was one of the pioneers with his ashram in South India and books such as *The Marriage of East and West* and *A New Vision of Reality.* He saw a unity underlying the expressions of the great religious traditions and used scriptures from these different sources in his own practice. Meditation workshops have played a key role in the conferences and meditation forms a key part of the daily practice of many conference participants. Ravi Ravindra has illuminated comparisons between yoga and physics, as well as writing from an Eastern point of view about the Gospel of John. As well as introducing his notion of 'creation spirituality,' Matthew Fox has encouraged a return to sources of Western mysticism such as Meister Eckhart and Hildegard von Bingen. Anne Baring has been part of a movement enhancing our awareness of the Divine Feminine and her role in the renewal of our culture. More broadly, there has been considerable debate about the relationship between religion and spirituality, with many people espousing the latter but remaining wary of the more formal and dogmatic aspects of established religions. Concomitantly, there has been an enormous expansion of new religious movements. Books on mystical and near-death experiences have, I believe, reinforced the trend towards more personal forms of spirituality.

Debates on the relationship between physics and mysticism are still very much alive. Ken Wilber, in his *Quantum Questions*, showed how many of the great physicists of the early part of the century had an interest in the interface and in the unity at the heart of the mystery of physical existence. I have already mentioned Fritjof Capra's *The Tao of Physics*, which was both widely acclaimed and criticized. He presents his case in a lecture included in this book. With the publication of *Wholeness and the Implicate Order*, the work of David Bohm began to receive widespread attention, especially his idea of a dynamic, underlying implicate order from which the explicate physical world arises. Bohm had been influenced by many years of contact with Krishnamurti and proved to be an inspiration to thinkers seeking to articulate a holistic world-view. It should be noted, however, that his deterministic interpretation of quantum theory receives little support from within the physics community, although the book published with Basil Hiley after his death, *The Undivided Universe*, has added to his standing.

Ideas of self-organization put forward in this book by Paul Davies have exerted an increasing influence on both physics and biology over the past twenty years. Ilya Prigogine and others have found that order may arise out of chaos and that his idea of dissipative structures, originally derived from chemistry, can be metaphorically carried over into other fields. A dissipative structure is an open system whose equilibrium is maintained through a relative constancy of input and output. When, however, the pressures become too great, the system spontaneously reorganizes itself at another level and re-establishes a new equilibrium. Such models have been applied to transformations on a number of different levels, from biological to social systems.

The most significant development in biology over the past twenty years is the emergence of the Gaia hypothesis, pioneered by James Lovelock and Lynn Margulis. Lovelock's original book was published in the late 1970s and has been followed by more elaborate presentations. The principal contention is that the biosphere is a self-organizing or self-regulating system. The implications of this view amount to a criticism of the neo-Darwinian picture of the environment exerting selection pressure on organisms 'from above.' The Gaian perspective sees organism and environment as one mutually influencing system and points to the importance of symbiosis and co-operation as forces complementary to competition and struggle for survival. This leads to the idea of synergy as the power of co-operation. Brian Goodwin has brought chaos and complexity theory to bear on biology and has been a pioneer in the reintroduction of Goethe's scientific ideas involving a participatory view of nature. These developments have occurred largely within the dissident stream of biology, while mainstream molecular biology has remained loyal to the central dogmas of neo-Darwinism.

Even more revolutionary have been the ideas of Rupert Sheldrake. Ever since *Nature* suggested in its review of his book *A New Science of Life* that it was a good candidate for burning, his ideas of morphic resonance and formative causation and his contention that the laws of nature are more like habits have been controversial. His basic idea that morphic fields make it easier for things to happen again if they have already occurred has been successfully tested in a number of ways. His more recent work has extended to a new philosophy of nature and to *Seven Experiments that Could Change the World.* Among the most popular of these have been his experiments on the sense of being stared at and his experiments with pets who know when their owners are coming home. Such work may be ignored by sceptics in the mainstream, but he is setting up genuine experiments and involving large numbers of people in research which is not only inexpensive but could also change the way we consider the mind. His idea of the 'extended mind' owes something to the metaphorical extension of quantum non-locality. Experiments have shown that a widely separated particle that was once joined to another responds immediately to a change of spin in the other.

I have already mentioned the recent explosion of interest in consciousness studies and the way in which this development was anticipated by the conference in the early 1980s. In an effort to become a respectable science, twentieth century psychology chose to follow the laboratory path of Wilhelm Wundt rather than the broader approach of William James, whose classic *Varieties of Religious Experience* is still widely read today. James was interested in integrating psychical research into psychology, but the emerging behaviourist school fiercely contested this move; they ensured that Harvard refused the offer of a chair in psychical research. The triumph of behaviourism marginalized psychical research, and later parapsychology. It has taken nearly a hundred years to return to the methodological starting point of William James as parapsychology gradually becomes an integrated aspect of consciousness studies. Charles Tart has been a pioneer in this process of integration, as is apparent from his piece in this book.

A recent obituary of Sir John Eccles describes him as a rare breed of neuroscientist in arguing that consciousness cannot be reduced to brain function. In the period since the delivery of his lecture, Eccles developed a theory of so-called psychons to account for the interaction between mind and brain. Interested readers can find this explained in his final book *How the Self Controls its Brain* (Springer Verlag).

Larry Dossey's work has contributed towards the emergence of mind-body medicine, a field that represents another aspect of consciousness research. Recent advances in psychoneuroimmunology have demonstrated the role of intentionality in healing and the importance of positive mental attitudes for the maintenance of health. Readers will find that many of the lectures printed here serve as introductions to the major developments I have outlined.

Experiment and Experience: Complementary Approaches to Truth

The words 'experiment' and 'experience' both derive from the Latin root *experiens,* meaning to try thoroughly. It is significant that the verb to experiment is now intransitive: one must experiment on something, thus distancing oneself from it, whereas the word 'experience' is direct; one experiences something at first hand. The suffix *periri* is related to the word peril, or trial to be passed through at some personal risk. Both experiment and experience may be dangerous. It is common to speak about the 'pursuit' of truth, a Promethean or Faustian striving that suggests something to be grasped and held. The etymology of the words 'comprehend' and 'concept' reinforces this idea, as they are both derived from *cum,* meaning with, plus *prehendere* or *capere* meaning to grasp. An alternative set of metaphors proposes that truth is seen rather than grasped: the origin of the word 'theory' goes back to the Greek word for contemplation *theoria,* from which the word theatre is also derived.

Then what is the purpose of apprehending truth? It is to accumulate information, gain knowledge, acquire wisdom or a combination of all three? I would contend that we are all seeking insight and understanding, which is surely one reason why we attend conferences like these! At the deepest level, I believe that truth has a transpersonal or impersonal aspect that is recognized by means of an inner intuitive faculty; hence Browning's suggestion in his poem *Paracelsus* that truth arises from within ourselves. Indeed, perhaps understanding itself is always intuitive although it may have a rational structure or justification.

If we now ask what an explanation is, we may find ourselves having to explain the explanation and consider the nature and level of causality involved. An explanation is unfolding of the implicit, a restatement of the issue in terms of assumptions regarded as ultimate. So explanations lead us back to assumptions that also dictate how satisfactory an explanation is perceived to be. For instance, current science tends to favour reductionist and evolutionary types of explanation. The next question is what counts as evidence? It may derive from experiment or perhaps from experience, which in turn will determine your method of validation. It is worth recalling Heisenberg's remark that experiments are a mode of questioning and the quotation from Niels Bohr: 'It is wrong to think that the task of physics is to find out how nature is. Physics concerns what we can *say* about nature.' What we say about nature, then, depends on our assumptions, intellectual framework, world-view and level of analysis.

If one accepts Aldous Huxley's contention of knowledge as a function of being, then one has to ask about one's level of knowing. St Bonaventure suggested that we have three eyes, the eye of sense, the eye of reason and the eye of contemplation. These could be called the outer, logical and inner eyes, or, in Plotinus' terms *nous* (intellect), *dianoia* (reason) and sense. He distinguishes between the first two by saying that: '*Nous* beholds things in their true relations beyond the separating and recombining function of discursive reason.' Reason gives us an analysis of the parts while intellect brings an insight into the whole. Natural science restricts its scope to the eyes of sense and reason. If a mystic is being tested in the laboratory, the instruments can only record changes in EEG patterns. The experiment cannot validate the inner nature of the subjective experience. In this spirit, Radhakrishnan wrote that: 'a physicist who rejects the testimony of saints and mystics is no better than a tone-deaf man deriding the power of music.'

In mystical insight the self or knower is not separate from objects of knowledge. As Seyyed Hossein Nasr asserts: 'knowledge has become separated from being and the bliss or ecstasy which characterizes the union of knowledge and being.' In Indian philosophy this is expressed as the union of *Sat-Chit-Ananda,* or being, consciousness, bliss. The process of the externalization of knowledge or the vision of duality 'blinded him to the primordial knowledge which lies at the heart of his intelligence.'

Nasr further comments that the divine source of knowledge is also source of being and love: hence their union in supreme knowledge. As he puts it: 'The very essence of things is God's knowledge of them and that there is a reciprocity and, finally, identity between knowing and being.'

Or Plotinus: 'It is not the eye which sees, but the active power of the soul. It is this power that is the root of understanding.'

Or Novalis: 'How do we see physically? No differently than we do in our consciousness — by means of the active power of imagination. Consciousness is the eye and ear, the sense for inner and outer meaning.'

We reach the heart of the paradox of knowing when we realize the non-duality of knowing like Eckhart: 'The eye with which I see God is the same eye with which God sees me.'

And Ruysbroeck: 'By means of this inborn light they are transfigured, and made one with that same light through which they see and which they see.'

If we ask at this point: What is real? then Plotinus would respond: 'We are within a reality that is also within us.' Immanence and transcendence are united in this level of knowing. Renée Weber observed a decade ago that 'a parallel principle drives both science and mysticism — the assumption that unity lies at the heart of our world and that it can be discovered and experienced by Man.' The universe is one, it is our perception that fragments it and alienates us from our intrinsic sense of belonging.

The spirit of science — the 'horizontal'

Many great creative scientists express aesthetic or mystical insights in connection with scientific knowledge. Einstein maintained that 'the cosmic religious feeling is the strongest and noblest motive for scientific research.' While the French mathematician Henri Poincaré wrote that 'the scientist does not study nature because it is useful; he studies it because he delights in it, and he delights in it because it is beautiful. If nature were not beautiful, it would not be worth knowing, and if nature were not worth knowing, life would not be worth living.'

This is much more than dry statistical analysis! The poet Wordsworth wrote that 'poetry is the breath and spirit of all finer knowledge; it is the impassioned expression which is the countenance of all science.'

I would like to propose that a similar spirit inspires the search for truth in both science and spirituality:

Exploration expressed in wonder and curiosity.
Creativity and imagination — bringing forth new models and discoveries
Critical and analytical rigour — applied to methods and procedures
Practice — through experimental prediction and testing
Openness and awareness of metaphysical assumptions

Science has been at its strongest in the first four areas, but awareness in the fifth domain of assumptions has often been lacking. Some great scientists, however, have been acutely aware of their importance, for instance Prince Louis de Broglie: 'History shows that the advances of science have always been frustrated by the tyrannical influences of certain preconceived notions that were turned into unassailable dogmas. For that reason alone, every scientist should periodically make a profound re-examination of his basic principles.' The fact that no philosophy or sociology of science is taught to the majority of science students does not encourage the kind of re-examination recommended by de Broglie, but the emerging science of consciousness may demand it. The co-originator of the theory of evolution, Alfred Russell Wallace warned that 'whenever scientific men of any age have denied the facts of investigators on *a priori* grounds, they have always been wrong.' Wallace himself was interested in psychical research and spiritualism, much to the dismay of his scientific contemporaries, but he knew that their prejudice was based on ignorance of the field.

Nearly a hundred years ago William James warned of the dangers of scientism, the conviction that only material world is real and only physical causation is scientifically respectable:

Science taken in its essence should stand only for a method and not for any special beliefs, yet as habitually taken by its votaries, Science has come to be identified with a certain fixed general belief, the belief that the deeper order of nature is mechanical exclusively, and that non-mechanical categories are irrational ways of conceiving and explaining even such a thing as human life.

At this point we should remind ourselves of the roots of the word science itself in the Latin *scientia*, meaning knowledge. We have seen that physical science (physics was originally called natural philosophy) is limited to the eyes of sense and reason. It therefore exceeds itself if it declares that what cannot be measured does not exist.

In his recent work Ken Wilber has proposed four quadrants of knowledge, all of which are necessary to give a complete account of reality. By distinguishing between individual and collective, interior and exterior, he arrives at these four categories of interior-individual (intentional), interior-collective (cultural), exterior-individual (behavioural) and exterior-collective (social). Science has concentrated mainly on the external or outside in perspectives and has tended to see the interior or subjective aspects of existence as secondary. The following chart amplifies this point. If one takes the direction of causality from the left to the right column, one would then conclude that the elements of the second column derive from those in the first. An alternative model, however, would propose that the factors are complementary rather than primary and secondary.

Causality and explanation

Primary	*Secondary [gives reductionist view]*
External/visible	Internal/invisible
Surface	Depth (shared)
Observation	Interpretation
Quantity	Quality
Objective	Subjective
Third person (it)	First person (I)
Matter-energy	Spirit-mind
Body/brain	Consciousness

These categories might in turn be expressed as:

Experiment	Experience
Science	Spirituality

In his book *Eye to Eye,* Wilber maintains that there are various valid modes of knowing, depending on the eye in question — sense, reason, contemplation. Each mode has its own injunction, mode of apprehension and method of confirmation. The *injunction* states: if you want to know this, do this. The procedure brings forth a particular data domain that may be an experiment or an experience. *Apprehension* comes in terms of a particular experience, realization or piece of evidence. *Confirmation* concerns testing and falsifiability, discrimination or interpretation and can equally be applied to different domains.

The science of the spirit — the 'vertical'

Any new science of consciousness or of the spirit requires what Wilber calls an integral approach using all four quadrants and noting their correlations. One of the differences between modern scientific as compared with inner spiritual research is that the researcher is trained but rarely transformed by the process of research. In a recent article for the *Journal of Consciousness Studies,* he notes that you can understand the writings of Daniel Dennett with an unreconstructed consciousness, but the same cannot be said for an understanding of Plotinus. Without an inner transformation you will think that Plotinus is 'seeing things' — 'and he is, and so could you and I if we both transform our perception and understanding.' Thus a profound science of consciousness or of the spirit requires an interior transformation of the researcher. Science itself can become a spiritual path to the extent that there is a change of vision and hence of the organs of perception. Objective perception can be achieved in both science and spirituality and, according to Ravi Ravindra, consists in 'freedom from oneself.'

Ravindra has written perceptively about such issues. He maintains that if being is divorced from knowing, it leads to a 'fragmentation of our sensibilities.' He adds that 'to the extent that philosophy and theology become scientific, God is reduced to a mental construct. Theology thus becomes a rational profession dealing with metaphysical systems, rather than a psycho-spiritual path for the transformation of the being of Man.' The mystic is necessarily transformed on the living spiritual path, but the temptation for the philosopher or theologian is to resort to abstractions, taking refuge in words and systems.

We can now apply the same scheme to the search for inner truth:

Exploration — the mystery of inner space.
Creativity and imagination — visionary insight and inspiration.
Critical or analytical rigour — precision and consistency in the use of terms, the development of discrimination.

Practice — commitment, discipline, meditation, prayer

Openness and awareness of metaphysical/theological/cultural assumptions. Tolerance is vital here.

Integration and the search for unity

The world picture of Plotinus sees the universe as a living chain of being, an unbroken series of ascending or descending (systole/diastole, in and outbreath) values and existences. The lower grades are less perfect and less real. There is an outward flow of creative energy that is balanced by a current that carries all back towards the source of being. This is the source of spiritual, intellectual, moral and aesthetic aspiration. In *Sex, Ecology, Spirituality,* Wilber makes the case that Plato united both the processes of descent and ascent, rather than being an otherworldly dualist. He recognized that the ground and goal of being are one and the same, that the world was the field of both descent into strife and ascent towards love. Wilber shows how Schelling arrived at the same conclusion and thus overcomes the dualism of spirit and matter. Schelling saw mind and nature as differentiated but also that the transcendental and unifying ground of both had been forgotten. He contended that Spirit is the very process of the One expressing itself through the many: mind and nature or matter are both grounded in Spirit. Thus Spirit is the only reality, beyond mind and nature: Spirit descends into manifestation, *but this manifestation is nevertheless Spirit itself, a form or expression of Spirit itself.* The importance of this insight is incalculable in the search to overcome the unreconciled dualistic categories of modern thought.

As Wilber concludes: 'Spirit knows itself objectively as *nature*; knows itself subjectively as *mind;* and knows itself absolutely as *Spirit* — the Source, the Summit, and the Eros of the entire sequence.'

He sees the process as a passage from subconsciousness through self-consciousness to non-dual superconscious; or from the pre-personal through the personal to the transpersonal; or again from the biosphere through the noosphere to the theosphere.

I believe that it is possible for science, and especially the science of consciousness to deepen into what Nasr calls *scientia sacra.* Here the mind is not divorced from the heart: 'the eye of the heart perceives the unity that is at once the origin and end of the multiplicity perceived by the mind and the mind's own power to analyse and know discursively.

The light with which the mind is seeking to discover is itself a ray of the Light of God.' We return to that earlier remark by Eckhart: 'The eye with which I see God is the same eye with which God sees me.' And Augustine: 'Our whole business of life is to restore the health of the eye of the heart whereby God may be seen.' The eye of the heart is precisely the eye of contemplation that can be opened. Our journey takes us from understanding to wisdom, from *theoria* to realized gnosis: 'the realization of a knowledge which being itself sacred, demands of a man all that he is. That is why it is not possible to attain this knowledge in any way except by being consumed by it.' Or as Rumi put it: 'The result of my life can be summarized in three words — I was immature, I matured and I was consumed.'

Cortona, Italy, September, 1997

Cosmology and Physics

The New Physics and
the Scientific Reality of our Time

FRITJOF CAPRA

Fritjof Capra took his Ph.D. at the University of Vienna and has done research into theoretical high-energy physics at the University of Paris, the University of California, Santa Cruz, Stanford University and the University of London. He is the author of The Tao of Physics, The Turning Point, Uncommon Wisdom *and* The Web of Life. *He is currently Director of the Berkeley based Center for Ecoliteracy, dedicated to promoting ecology and systems thinking in primary education through teacher training workshops, publications and a grant-giving programme. This lecture was delivered at the 1978* Mystics and Scientists *conference.*

What I would like to do here is to widen the horizon and to place the worldview of modern physics and its approach of the mystical traditions into a wider social and cultural context. If we look at our society and culture then we notice a basic philosophical and cultural imbalance. It is an imbalance between two modes of consciousness, which always been recognized as essential aspects of human nature and have traditionally been called the 'rational' and 'intuitive.' They have been associated traditionally with science and religion, respectively.

The Chinese have called these two modes of consciousness, or these two types of knowledge, the *yang* and the *yin*. They have never seen them as experiences belonging to separate categories, but rather as extreme parts of the single whole — as extreme poles of one and the same reality. In the Chinese view, all manifestations of reality, and this includes manifestations of the human mind, are generated by a dynamic interplay of these archetypal polar forces, the yang and the yin. To give you an idea of how the Chinese envisage this interplay, I will quote from an ancient Chinese text: 'The yang, having reached its climax, retreats in favour of the yin. The yin, having reached its climax, retreats in favour of the yang.' It is enlightening as a cyclical interplay between polar forces.

I think it is interesting and very instructive to observe the attitudes of our culture and our society with regard to these two polar aspects of human nature. The yang aspect is the masculine side, the active, rational, competitive, scientific side of human nature. The yin aspect is the feminine side of human nature. It is the yielding, intuitive, co-operative, mystical side. If you think of these two aspects then you will see immediately that our society has consistently favoured the yang over the yin. It has favoured activity over contemplation, rational knowledge over intuitive wisdom, science over religion, competition over co-operation and so on. Our culture takes pride in being scientific, it is dominated by rational thought and scientific knowledge is often considered the only valuable knowledge, the only acceptable kind of knowledge. That there can be other kinds of knowledge, an intuitive type of knowledge or awareness, which is just as valid and reliable, is generally not recognized. The intuitive mode is suppressed together with all the other yin or feminine aspects of human nature. Furthermore, instead of recognizing that the personality of each man and each woman comes about through an interplay of masculine and feminine elements, we have established a static and rigid order where all men are supposed to be masculine and all women supposed to be feminine. Together with emphasising the masculine, or yang, aspect our society, this has given men the leading roles and most of its privileges. However, we are now witnessing the beginning of a tremendous cultural change, which I think is an evolutionary movement. As the Chinese text says, the yang, having reached its extreme, is beginning to retreat in favour of the yin. We are witnessing a rising concern with ecology. We are witnessing a strong interest in mysticism, a strong interest in psychic phenomena, a rediscovery of holistic approaches to health and healing and, last but not least, a strong rising feminist awareness.

I think that all these movements and trends are part of the same evolutionary movement from an over-emphasis of yang qualities towards a balance between the yang and the yin. This is illustrated by the fact that modern physics, the shining example of science produced by an extreme specialization of the rational mind, is now making contact with mysticism, the essence of religion and the manifestation of an extreme specialization of the intuitive mind. This shows very beautifully the unity and complementary nature of the rational and intuitive modes of consciousness. The similarities between the views of physicists and mystics will therefore constitute a powerful argument in convincing our social and political institutions of the cultural imbalance that we are observing.

The emphasis on rational thought is epitomized in Descartes' celebrated phrase, *Cogito ergo sum:* 'I think therefore I exist.' This statement has led Western individuals to equate their identity with their mind instead of with the whole organism; 'I think therefore I exist' — not 'I feel,' or 'I dream,' or anything like this. 'I think therefore I exist.' And this attitude led to the separation of the mind and the body and to the Newtonian view of the Universe as a mechanical system consisting of separate objects. This Newtonian view has had a tremendous influence on all the other sciences. The natural sciences, as well as the social sciences and the humanities, have all modelled themselves after physics. Physics has always been the prime example of an exact science and what they have done is to model themselves after classical Newtonian physics. This is to adopt the mechanistic-reductionist view of Newtonian physics. Now that the physicists themselves have gone far beyond the Newtonian model, it is time for the other sciences to become aware of this development and to expand their underlying philosophies.

The applicability of the Newtonian model

I want now to discuss the impact of Newtonian physics and the Newtonian worldview on the other sciences, then the possibilities and implications of the New Physics. Before doing that, however, I have to make a point that I consider quite important. The new conception of the Universe that has emerged from modern physics does not mean that Newtonian physics is wrong, or to say that quantum theory, or relativity theory, or any of these newer theories are right. We have come to realize in modern science that whatever we see in a scientific context — all of our theories, our models, our concepts — is always limited and approximate. To put it in an extreme way, scientists are not concerned with truth, they are concerned with approximate descriptions of reality. What we, as scientists, do is construct such an

approximate theory, a model, and we improve the model by construct-
ing successive approximations. Each theory will be valid for a certain
range of phenomena and there it will be very useful. However, when
we go beyond this range of applicability to new phenomena we will
have to expand the existing theory, or replace it by a better one. In
either case, the theory will always be an approximation, it will never
have an absolute proof within the scientific context.

Now, this is of course very disappointing if you think that scientific
knowledge is the only kind of acceptable knowledge, but if you accept
other kinds of knowledge, it's not disappointing at all. In fact,
physicists who have really come to this belief and carry it out in their
theories are very optimistic and very balanced people, rather than
pessimistic and disappointed frustrated people. So, this is not at all a
pessimistic view, but its very important to realize that all scientific
theories and models are approximations. The question then will be,
how good an approximation is the Newtonian model as a basis for the
various other fields?

In physics itself the Newtonian model had to be abandoned when
scientists researched the realm of the very small — of atoms and
subatomic particles — and the realm of the very large — stars,
galaxies, clusters of galaxies and so on. However, it is very useful still
in what has been called the 'zone of the middle dimensions.' In the
other fields that I am going to talk about, biology, psychology and so
on, the limitations of the Newtonian model may not have to do
anything with dimensions, but may be of a different kind. What I am
talking about is not so much the application of Newtonian physics to
other phenomena, but rather the application of the mechanistic and
reductionist worldview that underlies Newtonian physics. It will be
necessary for each science to find out where the limitations of this
world view lies in their particular field.

Among the many concepts and activities that have been influenced
by the Newtonian world view — and that must change to be consistent
with the views of modern physics — I shall talk about those relating
to health, healing and health care. They cover a wide spectrum, from
biological and medical science, to psychology and psychotherapy,
health education and health policy, social science, economics and
politics. All these factors influence health and health care. I have
chosen the issue of health and health care because of its many facets
and immediate individual and social relevance.

In biology, the Newtonian view has led to the idea that a living organism could be regarded as a machine constructed from separate parts. Such a mechanistic biology was introduced by Descartes himself who talked about the 'animal as a machine.' This idea was then extended to the human organism and the mechanistic view has dominated the life sciences up to the present day. To give you just one example, Joseph Needham, in 1928, wrote the following: 'In science man is a machine for if he is not, then he is nothing at all.' Here we have got a very extreme view of this machine analogy. It suggests that a living organism can be understood by taking it into pieces and trying to put it together again from an understanding of its parts. This was a methodology of classical physics and this approach still constitutes the backbone of most biological thinking today. To give you another example, I will quote from a current American textbook on modern biology, which is very well regarded and is a standard textbook in biology departments: 'One of the acid tests of understanding an object is the ability to put it together from its component parts. Ultimately molecular biologists will attempt to subject their understanding of cell structure and function to this sort of test, by trying to synthesise a cell.' Here we have it spelled out in great detail: the aim of molecular biology is to take things apart, understand the parts and then put them together again.

This mechanical model of biology had a strong influence on Western medicine and has led to the formulation of what is known as 'the medical model.' Western medicine uses a Newtonian mechanistic model of the human body, regarding it as a machine that can be analysed in terms of its parts. Disease is seen as an outside entity that invades the body and afflicts a particular part and the role of a doctor is the role of intervention, to treat that particular part. This is done either physically through surgery, or chemically through drugs and different specialists are treating different parts.

To associate a particular illness with a definite part of the body is, of course, very useful in many cases. However, Western medicine has over-emphasised this reductionist approach and has developed its specialized disciplines to a point where doctors are no longer able to view disease as a disturbance of the whole organism and they are no longer able to treat it as such. What they do is treat a particular part of the body and this is generally done without taking into account how this part is interdependent with the other parts. In particular, it is not usually taking into account how this part of the body relates to mental and other psychic functions.

A further characteristic of the reductionist approach of our medicine is its over-emphasis on bacteria. In the nineteenth century, Pasteur observed bacteria in disease and concluded they were the cause of disease. Western medical therapy has been organized around this erroneous assumption for the past one hundred years. It has not been realized that most of the time bacteria are not the primary cause of disease but are merely symptomatic factors of an underlying biological disorder. Instead of looking for the underlying causes of the disorder, such as deficient nutrition, or lack of exercise, or excessive emotional stress and so on, Western medical therapy has concentrated on destroying the bacteria. In this way, it can suppress or alleviate the symptoms, but at the same time causes further damage to the organism. This drug approach to illness is, of course, encouraged and perpetuated by the pharmaceutical industry and in a wider context by our whole social and economic system. Therefore, to overcome it, we shall need not only a new philosophical and scientific model, but also social and political action. This brings about the social context and opens the social and political dimension into which you are inevitably drawn when you study health and healing from a holistic perspective. I shall come back to this a little later on.

Let me now go on to talk about psychology, which traditionally has dealt with the 'part' on the other side on the Cartesian division. You can see very strikingly how classical psychology, like classical physics, is very much based on the Cartesian division that Descartes said existed 'between mind and matter.' Ever since this division was proposed, psychologists have wondered how mind and matter interact with one another. Once you have them separated, it is very difficult to see how they interact with one another. I think that the current confusion about the role and nature of the mind, as compared with the role and nature of the brain, is a direct consequence of the Cartesian division. I am not saying that I have the answer to this puzzle. However, I am pointing out that, if we had proceeded to a different mode where we recognize the unity of mind and body, we would probably find it much easier to deal with this problem which plagues neurophysiologists and psychologists.*

* Eighteen years later, Capra did have an answer to this puzzle and published it in *The Web of Life,* 1996 [Editor].

Based on the Cartesian division, two approaches have been developed to study the mind. Behaviourism chose to study the effects of mind on matter by studying behaviour. To do so the methodology of classical physics was applied to this purpose. Living organisms were seen again as machines that react to external stimuli and this stimulus-response mechanism was modelled after Newtonian physics. It was the classical cause-and-effect mechanism, the sort of billiard ball-type notion of cause and effect, which was applied to the study of behaviour. Behaviourism is still the mainstream of academic psychology and it defends its approach by claiming it is the only 'scientific approach' to the study of behaviour. This claim is, I think, very significant. The reductionist approach of reducing behaviour to the stimulus and response of machines is identified with science, so the only science that behaviourists seem to know, or accept, is the reductionist, Newtonian type of science.

The other main school of classical psychology, of course, is psychoanalysis. Freud chose to study the mind itself through introspection. He did not deal with matter, so to speak, or with behaviour, but with the mind itself. Nevertheless, he wanted to develop a scientific psychology and to be scientific it had to be modelled after classical Newtonian physics, the only physics around at his time. Freud wrote: 'Animalists are, at bottom, incorrigible mechanists and materialists.' Like physicists, Freud searched for the basic building blocks of the human psyche. He focused on basic instincts and he postulated their ego and super ego. In this way, he described the psyche in terms of psychological structures and the forces between those structures. The ego, the *id*, and the super ego are seen as some kind of internal objects that are in conflict. The mechanisms and the machineries of the psyche are all driven by forces and these forces are modelled after the Newtonian picture of forces. Freud very consciously used concepts of Newtonian physics to describe the machineries of what he called 'the introcyclical apparatus.' His notions of drive and defence correspond very much to Newton's notions of an active and a reactive force, for example.

We can see when we study psychoanalysis that it is very much affected by the Cartesian split. First of all, the split led to this mechanical model of the psyche. The theory of psychotherapy is very much affected by the separation between mind and body; psychiatry neglects the body, just as most of medicine neglects the mind. In psychoanalytic practice, it is absolutely forbidden to ever touch the patient — you separate psychological problems from physical problems. The unity between mind and body is completely disregarded. Classical psychoanalysis, like classical physics, also rests on the assumption that the observer, that is the therapist, is separate from the phenomenon, that is the patient, and does not influence the data. It is interesting that psychoanalytic practice has gone beyond this — even Freud himself went beyond this in his practice — but the theory was based on the objective description of phenomena without disturbing them by observing them. Freud had to do that in order to model his theory after classical physics.

I have not talked about the role of the observer yet, but it is something that is essential in modern physics. You are aware, perhaps, of talk about electrons appearing as particles and waves and about the fact that how they appear depends on the kind of experiment you do with them — and that is up to you to decide! So, if you ask an electron, so to speak, a particle question it will give you a particle answer. If you ask it a wave question, it will give you a wave answer. The electron is neither a particle nor a wave, it is something beyond, but in its interaction with the observer it shows particle properties or wave properties. In other words, the electron does not have any absolute independent properties, it shows one or the other in its interaction with the observer. We can never talk about nature at this level without talking about ourselves. Now this, again, will be apparent in more modern theories of psychology and I will come back to this a little later.

You all know, of course, that the psychoanalytic approach is very useful in many instances, just as Newtonian physics is often extremely useful. The question will be: 'Where are its limits?' In physics, classical notions have to be abandoned or modified when we go to realms beyond our everyday experience, beyond the everyday physical environment. Similarly, we may expect that the notions of classical psychoanalysis will have to be modified when we investigate the more remote realms of the psyche that lie outside of everyday experience. This is particularly true for mystical or religious experiences, which psychologists now call transpersonal experiences. These experiences, which many people have spontaneously and which often bother people because they come overwhelmingly, and are usually very puzzling. These experiences had no room in psychoanalysis and patients were usually classified as being psychotic or being escapist, they contradicted all the main principles of classical science and classical psychoanalysis. I will attempt to show you that newer attempts of psychological theories and psychotherapies will have to incorporate these transpersonal experiences.

I shall now turn to the social sciences and I shall mainly focus on economics, although I find it difficult to distinguish between economical and political phenomena. Present day economics, like most other sciences, is fragmentary and productionist. It does not take the interconnectedness of things into account, but judges them uniquely in terms of money profit. All goods are treated equally without consideration of the many ways in which they are produced. It is not considered whether goods are human made or occur naturally, whether they are replaceable or irreplaceable. It is not considered how they are interrelated with all the rest of the environment and the rest of the Universe. All values are reduced to that of private profit making. Therefore, it is inherent in the present economic system to disregard our dependence on the natural world. Our economic system is profoundly anti-ecological. Such an attitude is in sharp contrast to the pre-literate cultures and also in contrast to the use of modern physics. Modern physics forces us to see the natural world as an organic whole in which all parts are interdependent. It is a whole that is in a state of dynamic balance. It is self-balancing, self-adjusting and self-cleansing. This is very unlike present day economics and technology, which recognize no self-limiting principle. Our economy depends on high production and high consumption and the faith in technological and economic growth has become central to our culture. All growth is seen as good. The idea that there can be pathological growth is not entertained. Our economic system is based on expansion, yet unlimited expansion on a finite earth can never lead to this system of dynamic balance that we observe in the natural environment. Therefore, in this subtle system of nature, our technology acts like a foreign body and, in fact, we are beginning to see signs of rejection.

To sum up the influence of the Newtonian-Cartesian world view — and I have been able only to touch on a few aspects — we can say that all scientists have been tremendously influenced by the mechanistic and reductionist view that underlies classical physics and the general Western way of thinking. The method of reducing complex phenomena to basic building blocks and of looking for the mechanism through which these interact has become so deeply ingrained in our culture that it has often been identified with science itself and with the scientific method. Views, concepts, or ideas that did not fit into this framework were either not taken seriously, as in the case of mysticism or paranormal phenomena, or they were considered as psychotic, as in the case of spontaneous transpersonal experiences. However, we are now witnessing this cultural change that I mentioned in the beginning. Physics, the shining example of the hard science, which has always been held up as a model for all the others, is now transcending the mechanistic and reductionist worldview. It is leading us to an organic, holistic and ecological view, similar to that of mystics, of psychics and of people with spontaneous transpersonal experiences. This transcendence will not only have a strong impact on the other sciences, but will also be therapeutic and culturally unified when all these various experiences and cultural manifestations are brought into a consistent framework.

Implications of the new view for health and healthcare

What is health? Health was defined by the World Health Organization in the late 1940's as 'a state of complete physical, mental and social well being, and not merely the absence of disease or infirmity.' I am not quite happy with this definition as 'a state of complete well-being' because health is not a state, it is a process. However, we can take it for the time being. The important thing here is that the World Health Organization has defined it as 'complete physical, mental and social well-being.' Now, in view of this definition of health, we must say that our society at present does not have a system of healthcare. It has something that is called a healthcare system, but this is restricted to medical care, that is to fighting existing diseases within the mechanistic and reductionist medical model. Out of the large network of phenomena that influence health, which is biological, psychological, social and environmental factors, the medical approach studies only a few physiological aspects. It is not health oriented but disease oriented and it uses destructive techniques — drugs, radiation and surgery — as its main methods of therapy. There is practically no preventive healthcare, except for some preventive medicine in the form of vaccination. I am not saying that all physicians deal with health in that way, there are many physicians who live in their lives the Hippocratic ideal of medicine and our somewhat nostalgic view of the old family doctor. However, the medical system — the medical research, medical education in universities, the way hospitals are run, the whole structure of the health establishment — has this reductionist approach. It deals only with a small aspect of health and therefore, not surprisingly, it affects health only very marginally. In fact, quite often the system itself is a health hazard, as has been shown by Ivan Illich in his book *Medical Nemesis.*

Now this mechanistic and reductionist approach to health and healthcare is incompatible with the world view of modern physics, which has come to see the cosmos as a web of interrelated and independent processes. The medical model is praised as being scientific, but it follows the views of an outdated Newtonian science. The question then will be: 'How can we take the views of modern science as a basis for a new scientific approach to health and healthcare?'

It turns out that, to do so, we need not break completely fresh ground, but can learn from models existing in other societies. I have emphasised that the worldview of modern physics is strongly related to the views of Eastern cultures, the cultures of India, China and Japan. These are cultures where notions of the human body and healthcare have always been an integral part of natural philosophy and spiritual disciplines. Therefore, the study of traditional Oriental medicine should be extremely useful for our purpose. In contrast to the mechanistic reductionist approach of Western medicine, the Chinese idea of the body has always been predominantly functional. The body is seen as an indivisible system of interrelated components. Chinese medicine is not so much interested in particular organs in the human body, rather the way these organs are interrelated. It is evident that this view of the body comes much closer to the modern physicist's view of the body, which has come to see the interrelations, interconnections and interactions as more fundamental. This similarity is enforced by the fact that Chinese medicine, like modern physics, has come to realize that this interrelated web it is studying is intrinsically dynamic. The ideal state of health is seen as a dynamic balance between all forces within and without the body, which are in a state of continual fluctuation — a fluctuation that is described in the Chinese tradition by the interplay of the archetypal poles, yin and yang. The mental and emotional state of the patient is crucial to this balance, because the functioning of mind and body are seen as just different aspects of one and the same system. Furthermore, the human organism is seen in relationship to the whole cosmos. To keep it in balance, it is important to consider the patient's environment. By that, the Chinese mean not only the physical environment, which influences the organism through nutrition, climate and so on, but also the emotional environment, the psychological environment, the social environment, the spiritual environment and so on. The ideal oriental doctor, therefore, would be a sage who knows how all the patterns in the universe work together. It is a man or a woman who treats each patient on an individual basis, whose diagnosis does not categorize the patient as having a specific disease, but would record, as fully as possible, the patient's total state of mind and body and its relationship to the physical and social environment. Now, it's easy to see that such a holistic, ecological approach is likely to emphasise preventive healthcare, and, in fact, it has always been a main emphasis of Chinese doctors to prevent their patients from losing their balance, from falling ill. On the other hand, if you have a fragmented approach, which does not take into account interactions between mind, body and environment, there is very little room for preventive healthcare.

I want to emphasise that I am not suggesting that we should replace our medicine by Chinese medicine. What I am suggesting is that we expand the narrow, fragmented, mechanistic approach of our medicine into a broader holistic and ecological approach, similar to that prevalent in the Eastern traditions.

Restoring the dynamic balance

I want now to develop with you an approach to health and healthcare that will be consistent with modern physics and with models like the Chinese. I shall start from the notion that the state of health is a state of dynamic balance between the various aspects of the organism and between the organism and its environment. The various aspects and manifestations of this balance will have to be explored and methods will have to be developed to recognize an imbalance before symptoms occur. The role of the healer or therapist will be to assist a person in maintaining or regaining his or her balance. An important aspect of this dynamic balance, which is also in the Chinese medical system, is the recognition that the healthy organism is in a state of homeostasis. That means that it has the tendency to go back into a balanced state once it has been disturbed — the organism has a natural tendency to remain in balance. When the balance is disturbed, the therapist merely has to trigger a response. The organism will know that an illness can only be cured by the organism's inherent healing power. The healer or therapist merely assists the healing forces and creates the most favourite conditions for the healing process to take place. This is, in fact, the original meaning of the word 'therapist,' which comes from the Greek word *therapuin*, meaning 'to attend.' So, a therapist is merely an attendant or assistant, the patients themselves do the healing.

When we explore the different aspects of this dynamic balance, it may be useful to identify three aspects of the human condition that are fundamentally interrelated, which I want to develop into a rough conceptual framework for this holistic approach. The three fundamentally interrelated aspects of the human condition are the body, the psyche — here I include the conscious mind, the emotions and the unconscious, both the personal unconscious and the transpersonal unconscious — and the environment, where I include the physical environment, the emotional, social, political, cultural and cosmic environment. The holistic approach to healthcare will consist in maintaining or restoring the dynamic balance between all these forces, and I shall call this holistic healthcare. (I don't like the word 'holistic' very much. It is very ugly and I hope we will be able to drop it eventually and we'll just talk about healthcare as what true healthcare should be). True healthcare will then mean people taking care of their health individually and socially as a community, with the help of therapists.

Three aspects of healthcare

First let us talk about individual healthcare. Central to individual healthcare is the recognition that the individual is responsible for keeping his or health in balance. As I have said, the therapist mainly assists the patient in doing so and checks various aspects of imbalance. This is in sharp contrast to present attitudes that delegate responsibility to the doctor and the drugs. The medical establishment and the pharmaceutical industry, of course, play an important role in encouraging current attitudes. They maintain the mystique that surrounds the medical profession by speaking to us only in Latin and by encouraging patients to rely increasingly on drugs. What I am saying here is gained from discussions with many people during and after lectures and many people in the medical professions with nurses, various therapists, various healers and also physicians, medical doctors.

This attitude of the health establishment has given rise to the public image of the human organism as that of a machine, which is prone to constant failure, unless supervised by doctors and treated with the help of drugs. The notion of the organism's natural healing power and its tendency to stay healthy is not communicated and that trust in one's own organism is not encouraged. The philosophical error underlying this view is, again, the Cartesian division and the resulting tendency of men and women to view their bodies as machines, to identify themselves with their mind rather than with their whole organism. This error is enforced by the strong influence of the pharmaceutical industry, which is not interested in keeping people healthy primarily, but is interested in selling its products.

What, then, is involved in individual healthcare? Well the health of human beings is determined above all by their behaviour, their food and the nature of their environment. These are all things on which we have some kind of influence. You can see that the major health problems in our culture are direct consequences of behaviour and living habits, for example, heart disease and strokes are related to diet, to cigarette smoking, to lack of exercise. Cancer is related to diet, to smoking and even more to the way we cope with stress. Liver disease is related to excessive use of alcohol and excessive drinking is also responsible for most of the deaths and injuries in car accidents. You can just keep on going down the list of major diseases. The individual has the power and should feel the responsibility to maintain his or her health by the observance of simple rules of behaviour relating to sleep, food, exercise, alcohol, smoking and so on — the kind of things that your mother or grandmother has always been telling you!

Why don't we do this? Why don't we take care of our health in this way? Well, first of all, there is human negligence and human laziness, but there are also other reasons, one of them being lack of knowledge due to inadequate education and due to the all-powerful force of advertising. Second, there is a public lack of interest in public health care due to the dominance of the present health establishment, which is based on acute high-cost hospital-oriented and drug-oriented medicine. We see, therefore, that individual responsibility has to be accompanied by social responsibility, and individual holistic healthcare by holistic health policy. We cannot expect the individual on his or her own to fight the tremendous conditioning by the drug industry, the educational system and the rest of the health establishment. Therefore, I think that holistic healthcare will not be successful without social and political action.

So, we come to health policy and social healthcare. I shall briefly outline the goals of the holistic health policy by first summarizing the influence of our reductionist, anti-ecological economy and technology on our health. On the one hand, the unlimited expansion of technology has created an unhealthy environment, which imposes excessive physical and psychological stress on the organism. On the other hand, the dependence of our society on continuing economic growth has created a situation where considerations of profit maximizing are put before all other considerations, and that includes the health of the consumers. As a consequence, most of the products produced in this economic system are a direct threat to our health. For example, the food industry with the unhealthy types of food that is produced and sold, and you can also mention working conditions, cars which are dangerous to drive, dangerous road systems. You can just go down the line of various things that are produced. The threat of buying these products is increased by the inability of the medical system to see the human organism in its full physical, psychological and social context and to deal with health in such a holistic way. Therefore a holistic health policy will have to be a two pronged approach. On the one hand, it will have to exert influence on our economy and technology to prevent health hazards from being created, from being generated. On the other hand, it will have to re-orient the existing system of medical care to expand it into a system of true healthcare.

Let me come to the third kind of healthcare, to a holistic therapy. I think that the first and most important aim of holistic therapy will be to make the patient aware of his or her imbalance, meaning that the patient's problems will have to be put into a wider context. This context can be physical context, nutrition, lifestyle and so on. It can be an emotional context of stress, interpersonal relationships. It can also be a political context, for example a patient belonging to a minority group that is suppressed in some way, which is an extremely stressful, poor situation. Of course, it can be a combination of all of these.

The recognition of the context of these problems is the recognition of the web of interrelated patterns. Recognition, in itself, is therapeutic, isolation of the problem is not. This latter aspect is particularly true for cancer, which has been compared to a nightmare — one which happens without any apparent cause about which you can do nothing and which will go on for you don't know for how long. This is because cancer has been taken out of the context. If you can incorporate it into the context then you can see how it is related to emotional stress, how it is related to lifestyles and so on. In this way, you will have hope of fighting this disease yourself by changing these conditions.

So, we can say that the purpose of holistic therapy, whether it is physical therapy or psychotherapy, will be to make the patient recognize the wider context of his or her problems. At the moment we have no conceptual framework underlying this kind of therapy, we have no single theory to ascribe and understand the various aspects of the human condition that are relevant to health. However, there are certain models and techniques and this of course includes the medical model, which can deal successfully with some aspects. Therefore, I want to propose an approach similar to the approach in physics, the bootstrap approach. It is an approach where you develop a number of interlocking models, a number of models that are mutually consistent but that involve different techniques and different approaches to different phenomena. These models should all reflect the basic interrelation of the phenomena and should also reflect their dynamic nature. It will be the role of the therapist or the healthcare professional to find out which model or which approach will be appropriate in a particular instance. In my view, a general practitioner should have therapeutic skills both at the physical and psychological level and these skills should enable him or her to cope with most of the problems encountered in primary care. Patients with more severe problems will then be referred to specialists, who may be specialists of physical therapy, or psychotherapy, or may be social workers.

Such a healthcare system is in strong contrast to the present system, which neglects primary care and produces specialists who will work at the physical or psychological level, but will very rarely consider the inter-relation of the mind and body. Physicians rely heavily on drugs and spend virtually no time on counselling and comforting their patients. Now many physicians, not only the old family doctors, but many young physicians who have come with different attitudes, try to do that. They try to spend time on counselling, on giving psychological advice, but they do so as pure amateurs, because they have not been taught to do so. They are taught, at least in my experience from my discussions with American young physicians, that all diseases have psychological factors, but they are not taught how to deal with those factors. They are not taught the psychological or psychotherapeutic skills, just as they are not taught about nutrition, for instance. I asked one physician friend of mine,'What did you learn in medical school about nutrition?' He said, 'I didn't learn anything because I was absent on that day!' That's a very typical response. So, the integration of mind and body of physical and psychotherapeutic techniques will be very important, and that brings me back to psychology.

Behaviourism and the mind

I have shown how classical psychology and classical psychotherapy were conceptually related to classical physics. The first to expand psychology into new realms was Carl Gustav Jung, who developed concepts that come much closer to those of modern physics than those of previous psychologies. Jung's notion of the collective unconscious provides a link between the individual and humanity as a whole — in fact, a link between the individual and the whole cosmos. This link cannot be studied within the mechanistic framework. Jung was well aware of the fact that we have to go beyond the rational approach to deal with the unconscious and penetrate into these new realms. The collective unconscious and its patterns, the archetypes, cannot be defined in a precise way. A further aspect of Jungian psychology, which goes beyond Freud, is Jung's notion of the psyche as a dynamic system characterized by the flow of psychic energy, which he calls the 'libido' between opposite poles. This notion comes very close to the emphasis of dynamic balance in modern physics and is also closely related to Chinese ideas about mind and body. Jung also placed great emphasis on the mutual interaction between therapist and patient and I would like to quote him: 'By no device can the treatment be anything but the product of mutual influence in which the whole being of the doctor as well as the patient plays its part.'

More recently, many psychologists have begun to study holistic psychological patterns, or gestalts, rather than the previously postulated psychic elements. The Gestalt approach recognizes that, just as the physical organism cannot be fully understood by taking it into pieces and analysing the parts, so the properties and functions of the psyche cannot be understood by reducing it to isolated elements. In fact, a fragmented view or a fragmented experience of the psyche is not only inadequate to understand its functions, it is often the very cause of pathology. Therefore, a holistic psychology should recognize the unity and inter-relation of the three aspects I have mentioned previously, the body, the psyche and the environment.

Well, what then will the new psychology look like? We have to remember again that the rational mind can comprehend reality only through a succession of limited and approximate models. Psychotherapy can go beyond the rational model, so if you want to do psychotherapy you can work from your unconscious, as Jung and others after him have emphasised — you don't necessarily need a precise rational model. However, it will be useful, also in this case, to have an approximate model of the psyche. And therefore the approach that I have suggested earlier may also be applied to psychology, the approach of having different conceptual models for different patients for different phenomena.

As various models in physics are valid for various dimensions, or for various phenomena, so psychological models will be valid for various modes of consciousness. For example, ordinary modes of consciousness can be described by behaviourism — a rough Newtonian chain of cause and effect will give the gross behaviour patterns. The finer we get, the more interactions and interconnections will have to be taken into account and the more complicated things will become. To give you a simple example, if you raise the temperature in this room continuously, then at some point I will take off my jacket because it will get too hot. You can say that is nothing but behaviour — the temperature goes up, the man takes off his jacket because it gets too hot. However, you can be a little more careful and ask, when did he actually take off his jacket? You can propose he was brought up in a certain culture, which says that, if give a lecture, you have to wear a suit. It is not done to take off your jacket in the middle of a lecture and therefore you will wait for a certain time until it really gets too hot. That moment will depend on your upbringing, on your background. It will also depend maybe on what you did last night, on your memories of the evening. You can go on and on, the finer realms of the psyche will all play a certain role. So, on a first rough approximation you can use the behaviourist approach, then you can use, say, the Freudian approach, followed by finer approaches.

Talking about these various approaches and models, several psychologists have found it useful to distinguish three levels of psychotherapy. These are the ego level, which you could call the Freudian level; the existential level, which turns out to be related to phenomena of death and birth; and the transpersonal level, where individual problems are seen in their cosmic context. Most day to day problems will be solved at the ego level, for which Freudian techniques may often be useful. Others may require work at the existential level and many of the basic existential problems can only be solved at the transpersonal level. This implies that, in order to make use of the therapeutic effect of transpersonal experience or transpersonal awareness, the therapist has to be familiar with that kind of experience and has to be able to transmit basic trust in this experience. This transmission of trust is done essentially non-verbally and it will depend on the therapist's own degree of enlightenment. Here we come to the borderline between the therapist and the guru, if you wish, or the therapist and the sage. The rational means of psychotherapy are useful as tools that have to be grounded in the therapist's own transpersonal experience.

Well, this brings me to the conclusion of my lecture. I have tried to show how a cultural and philosophical imbalance, the over-emphasis of yang/masculine values, has led to a reductionist and mechanistic view of reality. This model was at first extremely successful, but was then extended far beyond its limits of applicability to a point where it has helped to create a physical and social environment that has become a threat to our health and well-being; maybe even threaten our survival. I have discussed this cultural trend from a physicist's point of view. Classical Newtonian physics was a scientific basis of the reductionist mechanistic worldview and became the model for all other sciences. Twentieth century physics has shown the limitations of the Newtonian world view in the most dramatic way. It has transcended the mechanistic and reductionist conception of the Universe to adopt the holistic and ecological view, similar to the views of the mystics of all ages and traditions. I believe, therefore, that modern physics can play an important role in bringing about the social and cultural changes that are required, because it will be easier to convince our institutions of the necessity of these changes if we can give our arguments a scientific basis. In doing so, it will be important to recognize clearly how the various attitudes of scientists, physicians, economists, business people and so on reflect the same unbalanced yang-oriented world view that we have to overcome. It will be of no use to just attack, say, doctors for their reductionist approach to illness, to attack politicians for defending sexist or racist views, to attack drug companies for their excessive emphasis on profits to the detriment of people's health. We have to show that their attitudes reflect the attitudes of our whole culture. We have to show how they are based consciously, or unconsciously on the same erroneous philosophical views and we have to realize that we are all caught up in these views. Society or the system is not some evil people 'up there,' but consists of all of us and we all have to change.

Cosmos, Matter, Life and Consciousness

DAVID BOHM, FRS

Professor David Bohm took his B.Sc. degree at Pennsylvania State College and gained his Ph.D. at the University of California, Berkeley. Later he was Assistant Professor at Princeton working in Plasma, Theory of Metals, Quantum Mechanics and Elementary particles before becoming Professor at Universade de São Paulo, working on fundamentals of Quantum Theory and on basic philosophical questions in Physics. From 1955 he was Professor at the Technicon in Haifa, leaving to become Research Fellow at Bristol University. Until his retirement, he was Professor of Theoretical Physics at Birkbeck College, University of London. His books include Quantum Theory; Causality and Chance in Modern Physics; The Special Theory of Relativity; Wholeness and the Implicate Order; Unfolding Meaning; On Dialogue. *He died in 1992. This lecture was delivered at the 1983 Mystics and Scientists conference.*

Quantum field theory

One of the essential features of quantum theory applied to fields is the appearance of what is called zero-point energy: that is the energy of a field in empty space is not zero but has a certain minimum energy, its zero-point energy. One can analyse this field in terms of waves and each wave will have a zero-point energy. In addition to this minimum energy, each wave may carry extra energy. When the field only has its zero-point energy, we say that space is empty. When extra energy is present, it appears as a particle or as a photon in the case of light. If we apply the mathematical rules of the theory and add up the energy contained in all the waves in empty space, the result turns out to be infinite, simply because the number of waves is infinite. However there is reason to suggest that wavelengths cannot be shorter than the Planck length which is about 10^{-33} cm. In this case the total energy is finite but it is still very large. For example, calculation shows that the zero point energy in one cubic centimetre of empty space would be far greater than the energy that would come from annihilating all the known matter in the Universe. Thus matter sits on top of this vast zero-point energy and it is like a small ripple on top of a large ocean of energy. This ripple is quantized, having definite values of energy and momentum and moves freely just as a ripple moves freely through the background water. Thus all matter, large and small, moves freely through empty space as if there were no zero-point activity present. There is no way of directly revealing the presence of this energy by passing particles or light through space, although the presence of this energy does produce small changes in the energy levels of, for example, hydrogen. Thus our attempts to detect this zero-point energy in empty space always fails even though this energy is actually far greater, according to this theory, than anything which we use to probe it with.

The nature of space

In ancient times there were two opposite ideas of space. One was that space is empty, and the other was that space is full, it was a plenum. Indeed throughout the ages in science and philosophy the popularity of these two ideas has alternated. During the nineteenth century the idea that space is a plenum was common and it was called the ether. The reason for postulating the ether was that, in order to enable waves to move through space, it was necessary to fill space with some kind of material medium. Otherwise the waves would not propagate. With the advent of Einstein's special relativity, it is assumed that the ether must be replaced by a space that is empty. However the appearance of quantum field theory has, once again, led us back to the idea that space is full! In fact, that space is an immense sea of energy and that matter is merely a small ripple on this sea. These views give us different aspects of the nature of space.

Our present view of the cosmos is that it is evolving in time. We assume that the Universe as we know it began with a Big Bang, starting from some small point and spreading out to become what we now observe. What from our point of view is a Big Bang is, from the point of view of empty space, a ripple. That is, the entire Universe is a very tiny ripple on a background of emptiness, which is also fullness! Perhaps there could be many universes like our own. Or perhaps this whole theory is provisional; it only goes down a certain way and there might be other levels of reality beyond anything we are able to know so far. This would suggest that we are only scratching the surface of reality. Indeed this probably will always be the case: we can always extend our knowledge but however far we go, there is always a limit. There is always the unknown beyond it. It does not look as if knowledge is going to comprehend the ultimate totality of all that is — whatever that may be.

Cosmos and chaos

The word *cosmos* is a Greek word meaning, essentially, total order, whereas 'chaos' means the opposite of total order. All of this order is basically enfolded. In a previous discussion I suggested that there will be a hierarchy of implicate orders and I gave a video game as an example. Suppose you have a television screen on which there appear spaceships, which, for simplicity, are just spots of light, moving in a way determined by a programme and you interact with these spots of light by pressing a button to change the programme and play the game. This is an example of a system that has two implicate orders. Suppose the game could be programmed to produce waves that focus and became spaceships and then these spaceships could turn into waves. This would be a programme that would imitate quantum mechanics. Here the first implicate order is just what appears on the screen, while the second implicate order is the programme which organizes the image on screen. It is implicate in that any part of the screen may come from all over the programme or, more specifically, any part of the programme that deals with the image on the screen. Thus we can regard the images as being enfolded and unfolded. There is a third implicate order which is the person who is playing the game or the person who constructed the machine, and so on. You can imagine computer engineers who decide to make a third programme controlling the second and a fourth controlling the third and so on. The whole idea could go on indefinitely and one would then consider the possibility of an infinity of implicate orders of which we are abstracting a limited number of levels.

Let's use this analogy for physics. Here empty space can be regarded as the first implicate order, the field which is able to focus and make particles which then fall back into waves. These re-focus again and again so a particle is constantly condensing, evaporating, re-condensing and so on. This process occurs with very high frequency. We only see the focused particle and the rest of the process is not visible. The first implicate order organizes this, and we will assume that there is another implicate order that organizes the first implicate order. We will call this the super-implicate order. We propose that quantum mechanics works in this way — that is, there is a second level implicate order which organizes the first level. There could be a third organizing the second, but at present there is no evidence for this third level. However there is no reason to stop at that level. It is like the content of the book *One, Two, Three, Infinity:* if you once get to 3 then there's nothing to stop you going further.

In this way one can see how a process is organized into a multi-dimensional order. Even a computer programme can be organized in this multi-dimensional way, that is, the programme is not necessarily structured in terms of a three dimensional space, but is organized in a more complex way. This organization requires information that need not be structured in three dimensions, even though the programme produces and works on a three dimensional structure. To illustrate what I have in mind, let us return to non-relativistic classical physics. Even here the particles are not described in ordinary three dimensional space, rather a multi-dimensional configuration space. Again in a hologram, even though the image is produced in a three-dimensional space, the mathematics that describes the field (the light waves from which the hologram is constructed) is not described in a three dimensional space.

Multi-dimensional order and non-locality

I can begin with an analogy to show what I mean by a multi-dimensional order. If you take a rectangular fish tank containing a fish and you have two television cameras A and B with one of them at right angles to the other, so that they are looking at the fish from two different angles. You then have someone in the next room who doesn't know about this and is looking at two television images produced by A and by B. The observer sees what appears to be two fish whose movements are interrelated in such a way that every time one of them moves, the other one moves as well. Then suppose that somebody touched the fish with a tiny probe that you could see in image A, but you could not see in image B. Thus when the fish was touched, its image in B would move without appearing to be disturbed. You would say that touching the fish in A had caused the fish in B to move. You would then conclude that there was a non-local interaction between the two fish. The point here is that there is no non-local interaction in the three dimensional order, there is only the appearance of a non-local interaction in the two two-dimensional images. Thus the process is understood in a higher dimensional order — that is the two images are two-dimensional but the reality is three-dimensional. This is exactly the way the quantum mathematics describes non-local processes. Thus the interaction of two particles is actually described locally in a six dimensional space, not non-locally in two three-dimensional images which are images projected from the six dimensional space. It is these projections that give the appearance that when one particle moves, the other immediately moves in a correlated way, or when one particle is touched, the other one will move immediately.

This leads directly into what is called the Einstein, Podolsky and Rosen paradox where some property of one of the particles is measured, say A, then the corresponding property is induced in B immediately. In other words, when you do something to A, B is affected immediately even though it may be many meters from A. This effect was pointed out by Einstein, Podolsky and Rosen and used as a criticism of quantum mechanics. They claimed that it showed that there was something wrong with the theory; it was, in some sense, incomplete. But in fact actual experiments show that it is correct even though it violates the idea that connections should be local. However that problem disappears if the effect arises from a six-dimensional reality, that is from an entirely different order.

In ordinary conditions this sort of connection has a negligible effect, and we see all the particles in the same three-dimensional space, interacting through local interactions. However in a more subtle, careful analysis which is more accurate, you see effects which require this higher dimensional order. Therefore we say this three-dimensional space is an approximation to some higher dimensional order.

Something similar could be said about time too, but I won't discuss that for the present. Let me remind you that experiments have been done checking the Einstein, Podolsky and Rosen experiment. The latest were the Aspect experiments which check quantum mechanics under very stringent conditions. In particular they showed that you can do something to A and proved that there is an effect on B in a very short time, so short that there is no time for a light signal to pass from A to B. This suggests that there is some real connection between A and B, which was hitherto unknown in classical physics.

Thus far I have been considering the particles as separate, taking these as ripples and treating each ripple as a separate entity which is an abstraction, like taking waves and thinking of them as independently existing from the ocean. This way of thinking is correct only up to a point. Now let's go back to consider the ocean in which the ripples are oscillating, or are in some kind of motion. Now to understand the connection of fields in different parts of space, as I said before, we should introduce a further new kind of field implied by quantum theory that is not describable in classical physics at all. This is what I call a simple implicate order or a quantum formative field. The reason I call it a formative field is that it determines, not only the interrelationships of all the particles, but actually their formation. That is, the particles are constantly forming and dissolving from the background and there is in actual fact no permanent particle at all; there is simply a field that is determining the kind of particles that will form. This is similar, as I said, to the video game in which the programme in the second implicate order is determining the space ships that will form in the first implicate order. Indeed it could determine the form of anything in the first implicate order.

This is similar to Rupert Sheldrake's idea of formative fields, but I want to emphasise that it applies to the basic and general laws of physics and, for the moment, I'm not considering living beings. This is not an assumption I've just arbitrarily introduced, but it is actually implied by the present mathematical formulation of the laws of quantum theory, if you try to understand what they mean. This formative field has several basic new qualities. First of all it has itself to be described as a multi-dimensional order, actually it could be infinite dimensional order, which acts on the three-dimensional first-level field. Secondly its action is to connect and relate fields in different parts of space. This action doesn't necessarily fall off to a negligible value when the two regions concerned are far apart — this is what is responsible for the non-locality. It is the manifestation of the multi-dimensional character of the field, as in the fish tank analogy. Thirdly the formative field has to be seen as a single whole; its overall mathematical status does not depend on the first level field. Rather the first level field depends on it. So the way entities are formed in the first level and the way they are organized is determined by the second level. This applies in the case of the computer programme determining the relationships to space ships and whatever goes on the screen at the first level . So particles are now related in ways that depend on the state of the whole; even their very existence depends on the state of the whole. It is the whole that determines them to be relatively in-dependent or not as the case may be. It becomes possible, for example, for particles to be brought together in a whole in which they are interrelated in ways that cannot be analysed in terms of the parts. In a mechanistic theory, all sub-wholes derive from particles whose existence and relationship are assumed to be pre-determined, but here the very existence of the particles, as well as their interrelationships, derives from the larger whole. This is close to an organic point of view for the universe: it is rather like an organism where the very existence of the parts, as well as their interrelationship, will alter according to the state of the whole. For example, the information content in the brain may affect all the organs.

Now, we have two levels of implicate order: the ordinary fields in three dimensions and the formative multi-dimensional field which organizes these fields. All the order and structural features of the first level follow from the second. The first level is a sort of plastic background on which the second one can operate and produce almost anything. This suggests the possibility of a third order super formative field: the second order would be an implicate order organized by the third and so on to infinity. One direction of research would be to investigate this to see whether one could get new mathematics which would transcend the present quantum theory. This is a difficult question, but it is something some of us are working on. Just as the quantum mechanical order of reality is more subtle and holistic than the ordinary, so the next one will be still more subtle and still more holistic. We would say then that the things that have appeared to us so far are merely on the surface of the whole thing. As we go deeper, we are constantly revealing new features: in fact the whole of science is based on this process, and we can say that the thing which first ap-pears is superficial, merely the display of something deeper.

I claim that all science works that way, but this goes much further than most science thus far. You see the idea that this desk is a display of what are actually atoms moving around at great speeds and so on is a typical scientific mode of explanation: what appears is explained as a result of something that does not appear.

Now the cosmos, matter and space-time are also organized in this way. The ultimate principle is unfoldment and unfoldment generates the order of time which we might perhaps discuss. We would like to include ultimately time as well as space, that is to say time is enfolded as well as space, so that the ordinary time order is not self-contained, it is contained in some other order, it is determined and organized in some other order. We discussed last time that life could be looked at as an unfoldment of an order. In this way you can begin to see that the distinction of living and non-living matter is not really sharp. In fact the theory of evolution implies that there is no distinction and therefore it all must have been enfolded from the very start. The possibility of life is enfolded and the unfoldment of that possibility is essential; it does not arise from the superficial appearance of things but from something much deeper.

Psyche and soma

Now I want to develop things further to show a close similarity in what has been said about matter, animate and inanimate, especially with the infinity of levels of the implicate order, each one organizing the next. You see this is the organization of sub-wholes in their inter-relationships. Structure is essentially the order, arrangement, connection and organization of various elements, and the implicate order is finally the source of all this: all the structure which appears is unfolded from a structure which does not appear. We can regard this in turn as a structure that has appeared and which is unfolding from a yet deeper level and so on. In the first instance it does not appear, but in the very statement that we know about it, it has appeared at least to our thought.

Now I would like to begin discussing the mind. There is a term which has been used a great deal, namely, psychosomatic. *Psyche* means mind, while *soma* means body. The word, psychosomatic, signi-fies the usual way of talking about the interrelationship of body and mind, the physical and the mental. I want to suggest a generalization for soma that will mean the physical in general and not merely the body. Some people seem to find this idea difficult, but bear with me for a while as I try to explain. It is useful, at times, to play with words because words have a habit becoming very fixed in their meanings as we use them habitually and we think they can only mean this and nothing else. But in fact a word can implicitly have a more extensive meaning that it can unfold. The trouble with the word psychosomatic is that it suggests two distinct features, each existing in itself and by itself and these two features simply interact. There are two quite different entities, one called the *psyche,* the other called the *soma,* either existing inside each other or sitting side by side and somehow affecting each other. The term is dualistic in its very connotation.

I am going to introduce another term, which I call *soma-significant.* 'Soma,' as I said, is body, matter, or the physical, and 'significant' means meaning; the fact that I put a hyphen between them means they are not really different. If I hadn't put in a hyphen you might have thought that there are two different things, one called *soma* and the other called *significant* which somehow interact.

Now I want to say a few things about meaning, because meaning is crucial to what mind is, to what language is, to what thought is and to what action is. First of all I say meaning is a kind of action. For example, suppose you're walking in the dark and you see a shadow. If the shadow means the presence of an assailant, your whole heart begins to race and the adrenaline flows so that you are ready to act, ready to run or to fight or to freeze. As soon as you see the shadow means something else, your whole physical stance changes. In this example we see that the soma and significance are not really separate.

Another good example that I came across a situation in which you find what you regard as a precious jewel; you get very excited. But as soon as you discover it is synthetic, even though it almost indistinguishable from the real thing, your interest becomes greatly reduced. So your mental state depends on what the object means to you, either it is just a bit of matter or it is a precious jewel. Similarly some ink marks on a piece of paper may mean simply some ink on paper or it may mean a message of great significance. If you see a pattern on the television screen, it may mean moving patterns of light, or it may be a gripping drama of real life in which each somatic physical form has a meaning.

Now the meaning is apprehended at a more inward, subtle and

implicate level than the form itself. That is, you cannot easily get hold of the meaning of the form. The form itself is a meaning that means form. It means you are looking at it in the significance of its form only. So now the meaning itself may take a somatic form which has a yet more subtle meaning. For example, the writing on the paper or the print is somatic as ink on paper but at the same time it means something. The television image is a moving spot of light somatically but it means all sorts of things. So what you can do then is to consider the meaning and display it, then you can discuss the meaning of that meaning. Thus we can discuss the meaning of meanings and so on.

Intentionality and meaning

A great deal of our thought is the unfolding of many levels of meaning. In principle we can go as far as we please, even going to infinity. This is the first point. The second point is that meaning is, in the first instance, intention; I mean signifies that I intend to do something. The intention unfolds a certain total action. For example, the intention to speak unfolds a large number of words, which you do not actually choose one by one. You could not talk that way, you could not act that way, you could not carry out any action that way: the intention to act unfolds a whole action. Even people who have been studying neuro-physiology would concede that before some simple action takes place, some nervous state arises which must be associated to the intention to act. These intentions appear in the nervous system before your act. It may be a superficial intention or it may come from great depths of inwardness, but I am still talking about intention in general. Now, in general, I do not know what my intentions are, that is, what I really mean. I may discover this by my actions, which may contain unintended consequences leading me to say that I did not mean to do that. So the intention unfolds action along with the unintended consequences. In acting, the full meaning of your intentions is revealed, but this would be very difficult, costly and destructive if you depended on carrying out all your intentions to find out what your intentions were. You could do all sorts of things you would rather not do.

What we can do in the next stage is to display the intention in many ways, for instance through the imagination. Imagination as an image in the mind which may be a visual image, or a tactile image or an auditory image or even some other sort of image. Einstein used to say that he didn't have his most important thoughts in terms of visual images, but in terms of how it felt. On the other hand, image is any form of display that may not be an image in the sense of imitating anything; it merely displays the option, that is, the intention. It is just as if you were watching all these programmes on hospitals where you see the oscillograph in the room displaying the heart beat. You see a pattern on the oscillograph that has no sense of the image of anything in the heart — it is merely displaying some feature of the activity which is significant, which has meaning. That's what I mean by display.

Now images can display, but display could take place in a much more subtle form which wasn't imitating anything, because image according to its root is an imitation of the appearance of something. The word 'display' has the same root as unfold, *displicare*, to unfold. The purpose of the unfoldment is to reveal something other than itself. In an action, you unfold an intention, but the purpose is to produce that action. However in a display, the purpose of the unfoldment is to reveal what is behind it. A display is everything laid out so as to show something clearly, but the display itself is not what is interesting, but it is what it means that is important. The well known psychologist Piaget studied and observed the growth of intelligent perception in infants and young children very carefully and these led him to say that perception itself flows out of a deep initial intention to act towards the object and to incorporate or assimilate it into a cycle of outward or inward activity. In the case of tactile perception you can see this very clearly: if you are handling something which you cannot see, you must move it around in your hand and from many subtle movements before you can get an impression of what the object is.

Scientists who have studied optical perception show that the eye is constantly moving in many ways. If this movement is stopped, the vision is either impeded or vanishes; it is not possible to perceive without an activity towards the object to be perceived, and that movement must have an implicit or explicit intention. Piaget concluded that a child has a deep intention to act towards the world. As the action comes back in the form of a sensation, he can then see to what extent the act was realized. Every action contains an implicit image of the object, an implicit structure of the object, structured according to the object that is expected. If the action does not meet exactly the object as expected then what comes back is a difference between the object

expected and the object as it is. From that difference you either change
the intention or change the object, and in that way they eventually
come to agree with each other. In that case you see that you have a
perception of the object. So there is a constant correction and a constant going out and having to come back and change according to what
is seen and so on. Thus perception is constant movement backward
and forward, inward and outward.

Levels of perception and thinking

I now want to suggest that this is not only going on in the way Piaget
describes in regard to external objects, but it goes on inwardly as well,
that is, in the nervous system itself and so on. Within the nervous
system, each level of activity is somatic in some sense. It may be very
subtle, it may be deep in the implicate order, but we can regard it as
somatic, and therefore it has a meaning. This activity may be physical,
chemical, electrical, quantum mechanical, super quantum mechanical,
but we would say it is some kind of soma, with different degrees of
enfoldment. But this somatic form has significance or meaning for an
inward level which, therefore, has an intention towards the form in
question. So there is no meaning that is without intention. Meaning
arises out of intention of some kind whether it is the intention to pro-
ceed, the intention to know or the intention to do. The intention need
not be conscious of course, it may be inbuilt, or even instinctive. The
intention is an inseparable involvement of action, it folds outward
toward the somatic form. It then comes back as a display of the
intention and its fit or non-fit would be whatever this object is,
whether it be inward or outward. Now in principle this could go on to
infinitely deep levels of implicate order. If the significance comes from
memory, however, it should probably be limited to some finite depth
of inwardness because memory seems to be limited both in scope and
subtlety of its content. We cannot remember the complex subtlety of
an actual experience, mostly it's a simplified image. The more often
you bring it back, the more simplified it gets and often it is discarded.

Is there any finite structure of meaning that comes from memory?
Memory will contain meaning because memory contains intention,
memory contains the way of approaching the object as has been
demonstrated in many scientific experiments. Watch how the eye
approaches an object and scans it in a certain way; once it learns to

scan in a certain way, it scans similar objects again and again in the same way so you use memory as an aid to perception in order to recognize an object quickly. This however can go very wrong if you look at something different or very subtle which will be missed by just scanning it in the old pattern.

Any finite structure meaning that comes from memory can deal only with a limited range of facts, as I suggested. Beyond this it breaks down and genuine perception goes beyond to that which is potentially infinite in the depth of inwardness. Unless it is infinite it is going to be limited and bound by some structure which is inadequate. This will lead eventually to something wrong which cannot be corrected within that limited frame.

Now as I said before, in the implicate order each level organizes the somatic forms that are more outward — you can see this happening in ordinary experience. For example, we organize material objects according to what they mean to us: a forest may mean something to paint for an artist, a source of lumber for a businessman, or something to walk through for somebody else — so the whole thing is organized differently according to what it means. If it means a source of money then you may cut it down or let it grow and keep it growing — it may become a very different forest or no forest at all. Normally the meaning of material objects depends on thoughts as well as habits, on our physical habits of various kinds which are connected with them; these are, in some sense, more subtle than objects, at least as known to us; they are not more subtle than the deep structure of the object, but they are more subtle than the object as we actually deal with it.

We can organize these thoughts with other thoughts that are more general, more subtle and more powerful in the sense of necessity and we can go on indefinitely. Ultimately past knowledge cannot serve here: we have to perceive new meanings and encompass them in a greater whole. This is creative perception, perceiving a new meaning. It is crucial for order, because if we try to stick to the old meanings based on memory, they are bound to be inadequate at some stage; and if you cannot see anything new it is just going to get worse and worse. This requires freedom from memory, going outward with exploratory new intentions and responding to what actually happens in this action, sensitively and perceptively. In order to see something, according to Piaget, you have got to explore it with some intention. This may be actual physical exploration or it may be only inward mental exploration, but it is actually an action whether it is inward or outward. It is an exploratory action with sensitivity and perception to see how it fits. Ultimately I say that soma-significance has to be able to go beyond all the limits and to be infinite. I would say that this is intelligence beyond what I would call mere skill and thought. As you think you develop skill and thought. As you ride a bicycle, you develop skill in bicycle riding. Whatever you do you develop skill in doing it, if you do it right. But that is always limited. Skill in bicycle riding will not help you too much in driving your car, or writing or speaking another language.

Therefore, skill in thought is also limited, because there is skill in certain kinds of thought you're used to. As a mathematician, you are used to mathematical thought, and you're very skilful at it, but if you have to think about something else, you might find it very hard. The other fellow could not think about mathematics because he is not skilful at it. What I am saying is that intelligence goes beyond skill and thought and has this principle of exploring the meaning and being sensitive to it. I think that this is general, it is not a particular skill. Any particular skill and thought can ultimately be imitated by a computer, but I don't think that intelligence in the way I've defined it here could be imitated because every computer system is going to be finite, however large it may be. Either you would have to have one computer controlling all those below it, in which case if there were anything wrong with that computer, it would be too bad; the other suggestion is that the computers would somehow correct each other, but that might lead to confusion because they might all be programmed in a similar incorrect way. People who work in computers are very enthusiastic about the idea of artificial intelligence being duplicated by computers, and they propose that the way they work is that if one computer makes a mistake, another one will correct it. But the trouble is that all the computers are programmed by human beings — they all share the same mistakes as well as being limited and finite. Similarly, human beings who act like computers will be unable to correct their mistakes; if everybody is programmed in the same way, we'll all make the same mistakes and never correct them. That's what often happens.

Order and organization

So I maintain that intelligent life is organized according to an infinite hierarchy of implicate orders. This is a proposal, I don't know that is so, but I'm saying it is a proposal to be explored: this is the meaning that I propose for this situation that we're confronting. We can explore it and see what happens, that is the basic mode of perception and intelligence. We do not begin with presuppositions, which have their place; if you have something new you don't begin with a presupposition that you take for granted, but rather you propose it for exploration. You know that intelligent life is organized according to this infinite hierarchy operating as soma-significance; I say that there is something similar but also different in matter. The similarity is that in the higher implicate order the formative field organizes the lower one. Of course, we had only two levels in quantum theory, but I suggested perhaps that may be only temporary. We may eventually discover more levels, and if we did, this would make the analogy much better. Moreover, it is clear that in soma-significance not only does the meaning affect the object — the activity of the meaning — but the intention of the meaning affects the object. The meaning is also altered by the state of the object.

We don't seem to have grasped that clearly in quantum mechanics yet, so that might be another way in which you could try to develop quantum theory. To make the analogy closer, perhaps there really is a two-way flow, but in the domains thus far investigated in quantum mechanics, only one way has been important. This might require a further extension of quantum mechanics. If this proposal is right then the analogy between mind and physical process is quite close. In this case we could look at evolution in a new way. According to modern scientific ideas, everything has evolved from the Big Bang where the universe started. When the universe started there were no such things as electrons, protons, neutrons or any particles of light or anything, something which would be totally indescribable and beyond the theory. As the universe expanded all these things began to form and gradually the whole thing unfolded. Therefore you have to say that, according to the modern scientific view which is generally accepted, everything we know has evolved from something else, including the most fundamental particles that are supposed to exist.

If I take this idea of space as full, that the vacuum is full of energy, then we could say this energy has evolved into various forms through various formative fields. So let's look at evolution, not only the evolution of life, but the evolution of matter in general and say that matter as a totality is a kind of super intelligence that has inward depth, that this intelligence created our bodies which are material and that our own intelligence is a further development of this super intelligence of matter: it is its totality. We ourselves could never create anything like ourselves and I don't think any of us would hope to do so in the foreseeable future. Therefore, it must be more intelligent than we are at the moment. Of course, it produced the whole universe and what might lie beyond it. What I mean by 'beyond the universe' is beyond the universe defined as that which has unfolded from the Big Bang. The word 'universe' is a bit ambiguous, it literally means 'to turn towards the one,' so the universe should really be a way of looking and not a thing. You are looking universally, or you are looking particularly. The idea would be that the turning towards the one or turning towards the many shall ultimately be the same reality as turning toward the one. That is, the many and the one are one. There are two views of one.

I want to extend this notion of soma-significance to the whole universe. Everything is somatic, that is, physical at its own level, which may however be very subtle and yet it has meaning for what lies beyond, or inwardly. This would suggest that the universe could be regarded as the body of the universal mind, just as the human being's body is the body of a particular mind — that is a proposal, I'm not saying that I know this to be the truth. So it is something you can explore that would become part of your intention towards the world to see what would happen if you moved in that way. So any meanings as necessary when taken seriously are an implicit intention, every intention is already a meaning. If we take only a finite number of levels, then we'd say the bottom level must be purely somatic while the top level is purely significant and mental, but if we think of the unlimited and infinite totality, there is no top and no bottom. They may even go round in a circle or spiral but the simplest idea is just to stretch on indefinitely. Each element is simultaneously created and sustained by the activity of the energy that unfolds this meaning and, at the same time, it carries a certain meaning in itself, part of the total meaning. So in soma-significance, the physical and the mental are two aspects of one movement, of unfoldment and enfoldment: they are not really different.

If you were to take the spaceship model, you would regard the spaceship as somatic and the meaning would be the computer programme. But then the computer programme is somatic and what organizes it is significant — that is the state of the nervous system of the computer programmer, which is also somatic. Something deeper is organizing that, and so on. Therefore there is no division between physical and mental, between living and not living and so on. The physical substance and meaning are apprehended in two different attitudes, intentionalities or dispositions of the mind. Your attitude is really an intention, a general intention to approach in a certain way or to act in a certain way, and this also enthuses the senses, it comes out in the observing instruments that we use in science. According to our intention we produce various kinds of observing instruments according to what the world means to us. We would not attempt to use certain kinds of instruments to try to observe an electron, because the word means to us that it is a small particle. A few hundred years ago nobody would have had that intention at all.

You may take an analogy like the two eyes in stereoscopic vision. If you have the old stereoscope there are two two-dimensional pictures, each slightly different. When you look through it, you suddenly see a three dimensional world. Obviously there is no three dimensional world you're looking at — they are only two-dimensional pictures, one on one eye and one on the other, but when they get into the brain, the mind organizes them as a three-dimensional reality. That is the way it is organized when you look through the stereoscope. There is a slightly different picture on each eye and this slightly different image helps to organize the sense of the three-dimensionality of the world that we see. So we say that the three-dimensionality that we see is evidently a construction of the brain; you can see it two-dimensionally or three-dimensionally. So we could say now that a similar principle may well hold more generally. That is, we could say that we have the soma as one way of looking at it, the significance as another way of looking at it; there are two, but both cover the same reality in the infinite, but in the finite, each covers a different aspect of reality, a variable aspect according to the way you're doing it.

The ground of soma-significance

In the whole, soma and significance are the same, but reality is neither soma nor significance because it is beyond both. It transcends both, in the same way that the three-dimensional world is neither one of the cardboard pictures nor the other that you see in this stereoscope. They are both, as it were, subsumed in this three-dimensional reality, they are both held in it as aspects so they have an entirely different meaning. Once you see the thing three-dimensionally, the whole thing means something entirely different. Therefore your intention towards it is going to be different. We do not choose our intentions. Freedom does not consist of the ability to choose your intentions because there is no way to choose your intentions. Intentions come because of what everything means to you, they are implicit.

Suppose we say soma is one side of reality, significance is the other. The reality of their sides cannot be described, at least not at the present stage, but it is our ground that we're describing — we have to see ourselves as within that reality that we want to describe. It would be a mistake to say we are thinking of a nice picture of soma and significance in front of us, as if we are outside it describing it. What is implied in this is that the beings that we think we are would become like the cardboard pictures when we see the thing three-dimensionally in the stereoscope. With our abstractions these are rather small things, and so the physical nature of the soma is one stream, mind-meaning is the other, intimately related because they are two sides of one reality. We do not reduce one side to the other, nor are they identical. What is identical is the reality in which mind and matter are the two sides. They indicate identity, but they are not themselves an identity. The one reality is not only identical, but it is simultaneously the potentiality, the ground of all difference and identity. This kind of identity is implicit difference, enfolded difference, where all differences enfold the identity from which they have emerged and to which they must return. That is, this identity which is the whole of all that is. It enfolds all the differences, it is not an identity without difference but contains all difference. All difference is implicitly that identity from which it emerges and to which it must return.

This cannot be grasped in terms of formal logic, which calls for saying either identity or difference. In terms of formal logic something must be either identical or different, they are mutually exclusive and contradictory terms. Formal logic is an abstraction from reason, but we should not hold to these fixed categories. It flows freely. We may then fix it for certain purposes, but we have to be careful not to hold it too fixed, that is it is only a useful device to fix it for the sake of discussion and to draw certain conclusions, but nowhere can you hold the certain categories fixed when we come to any large scope of discussion — you have to have what is called the 'unity of opposites.' This unity can only be put formally in the ordinary state of consciousness, but probably to be seen fully it would require another state in which the mind is organized in a higher and more subtle level of implicate order.

I could give an example of how the mind is organized three-dimensionally, or two-dimensionally as in the case of optical vision. There is a statue of Athene, the goddess of wisdom, in the Louvre in Paris in which one eye is directed outward reflectively into the distance, while the other eye is directed inward reflectively, so as you walk around you see the two eyes are: one outward, the other inward. Now presumably it is not that the outward eye is looking at the scenery and the inward eye at various personal concerns, but rather they are both looking at the same thing, that is the same reality. The outward eye and the inward eye are seeing the same reality, this is the essence of wisdom. As long as we see those two as different realities then we cannot, according to that statue, have true wisdom. So we could say that soma is the outward view of reality, significance is the inward view of reality, but they are the same in a higher reality.

Unification of outer and inner

Sir George Trevelyan has said that the West has emphasised the outward view of nature, the East the inward view of the mind. Both are probably one sided, and both have demonstrated a certain kind of inadequacy. The West's inadequacy is now becoming painfully obvious as it threatens to destroy the world. The inadequacy of the East is more subtle in the sense that things became stagnant there and when they were exposed to the West, they did not go on with their own culture but adopted the West's which is a rather mechanical thing to do, I think. They adopted some of the worst features, so that what seems to be required now is to make a new step in which we actually see the new meaning of both sides together. The difficulty of saying 'look only at the inward side' is this, that meanwhile the outward side continues on its own and the scientists develop whatever they do. Then there will be people who will be compelled to look at the outward side, but this is basically a mechanistic superficial approach which looks good in the beginning but leads to trouble in the long run. The inward side seems not to make contact with it. You may say inwardly that I look at all this and outwardly accept all the assumptions that are now current in the world and in science. However, this means that we're teaching fragmentation in one area while we're trying to deny it in another. What won't work is if most of our life is spent doing fragmentary thinking, then at the same time we try somehow not to do it — but this is not really going to be effective on the whole!

Either you must go into some retreat and look at things inwardly, or you try to engage with what is actually going on in society. Either way you will find a hopeless contradiction so it is necessary to bring these two together; the outward and the inward must be compatible, they must agree. There is a pressure towards that, because everybody generally throughout our society has accepted the mechanistic outward view so that the inward has become more and more mechanical. And you can see the effects by watching television programmes. Its effect must be to make people more and more mechanical. It seems therefore that the world view, the view you take of science and the view of the world as a whole, the world of matter, the world of physical reality and so on is important for the other side. We really have to bring it all together and therefore work on both sides seems to be necessary, which is really the point I wanted to make.

The Cosmic Blueprint:
Self-organizing Principles of Matter
and Energy

PAUL DAVIES

Professor Paul Davies is Professor of Natural Philosophy in the University of Adelaide. Previously, he held the Chair of Theoretical Physics at Newcastle University, gained his Ph.D. in London where he spent eight years as lecturer in Mathematics. His research has been in the area of quantum field theory in curved spacetime and he has investigated a number of quantum effects associated with black holes and the very early stages of the big bang. He is concerned with philosophical problems in modern physics, and the possibility of a completely unified theory of fundamental physics. He has been a consultant to Nature *and* The Economist *and is an academic editor for Cambridge University Press. His popular books include* The Runaway Universe; Other Worlds; The Edge of Infinity; God and the New Physics; Superforce; The Cosmic Blueprint; The Matter Myth; The Mind of God; *and* About Time. *This lecture was delivered at the 1988 Mystics and Scientists conference.*

In his book *The Phenomenon of Man* Teilhard de Chardin writes the following:

> After allowing itself to be captivated in excess by the
> charms of analysis to the extent of falling into illusion,
> modern thought is at last getting used to the idea of the
> creative value of synthesis in evolution. It is beginning to
> see that there is definitely more in the molecule than in the
> atom, more in the cell than in the molecule, more in society

than in the individual and more in the mathematical
construction than in calculations and theorems. We are now
inclined to admit that in each further degree of combination,
something which is irreducible to isolated elements emerges in
a new order. (p.268)

'Something in the cosmos,' he goes on, 'escapes from entropy and
does so more and more.' When these words were written, science still
had a strongly reductionistic flavour. The sentiments they expressed of
an unfolding, progressive, creative cosmos, though not especially new
to philosophy or theology, were nevertheless relatively heretical as far
as physics, cosmology and biology are concerned. Today, however,
following the rise of what we call the 'new physics' and the 'new
cosmology,' they no longer seem so heretical, and this evening I
should like to argue that Teilhard's belief in a creative progressive
cosmos has at last been vindicated.

The creative cosmos and the second law of thermodynamics

To understand why this image of a progressive cosmos was anathema
to scientists, one has to turn the clock back to the year 1857 to the
German physicist Hermann von Helmholtz. He proclaimed on the basis
of his so-called second law of thermodynamics — I say '*his* so-called,'
because there were many contributors — that the universe is dying.
(He didn't put it quite that way, but that is what he meant). According
to this second law of thermodynamics, the entire cosmos is engaged in
a one-way slide into a state of total disorder, popularly referred to as
the 'heat death.'

The prediction of a heat death profoundly influenced and indeed
depressed a whole generation of philosophers and scientists. Typical
of these was Bertrand Russell who wrote in his book *Why I am not a
Christian* in 1903, the following words in response to this dismal
prediction:

All the labours of the ages, all the devotion, all the inspiration,
all the noonday brightness of human genius are destined to
extinction in the vast death of the solar system, and the whole
temple of man's achievement must inevitably be buried
beneath the debris of an universe in ruins. Only within the

scaffolding of these truths, only in the firm foundation of
unyielding despair, can the soul's habitation be safely built.

As you can see, he was pretty fed up at the whole prospect!

Let me say a few words about this 'dismal' second law of thermodynamics. I'll do this with the aid of a very simple and familiar
example: Humpty Dumpty. He succeeded in disordering himself really
rather dramatically, in fact so dramatically that even with the attention
of all the King's horses and men he couldn't be reconstituted. This is
an example of what we physicists call an irreversible change. The
second law of thermodynamics is founded on the study of irreversible
changes and we can find many such examples in the world about us,
examples of where physical systems make transitions from one state
to another, but can't get themselves back again into the original state.
If one wants to summarize what these irreversible changes amount to,
it is that the later states tend to be more disordered than the earlier
ones. We can quantify this by saying they have a 'higher entropy.' If
this is applied to the universe as a whole, we are led to the statement
that Helmholtz made: that every day the universe becomes a little more
chaotic. Well, we know that from experience! The point is, though,
that this is a one-way, in-time process which imprints upon the world,
indeed on the whole universe, a so-called 'arrow of time.' It is the way
in which we distinguish past from future and it is very distinctive
because, in fact, almost all of the phenomena that we observe in the
world have this one-way-in-time character. Applied to the universe as
a whole it means that the cosmos, overall, is running down or
degenerating.

One way of looking at this is in terms of what is called 'useful
energy.' Now, it is often said there is an energy crisis, but most people
don't realize that energy is conserved, you can't get rid of it. Rather,
the problem is that energy changes from one form to another and you
lose control over it. So, the 'crisis' being discussed is not about energy
as such, but 'useful' or 'controllable' energy. The second law of
thermodynamics states that such energy gradually becomes dissipated
or dispersed around the universe and is no longer useful. Thus, it is
a degeneration of a stock of useful energy and the entire universe is
engaged in this degeneration. Every time anything macroscopic
happens in the universe, it degrades some energy and so leads
eventually to the heat death, in which there is no useful energy left —

everything essentially grinds to a halt, and nothing further of interest occurs. This was the gloomy, but apparently rock-solid, prediction of the science of thermodynamics, which has set the scene for all modern cosmological ideas. The universe may have had a birth, in what today we call the Big Bang, but it is also degenerating and eventually it will have some sort of death. This was the point that was influencing Bertrand Russell.

Alternative universes

The science of thermodynamics was developed in the nineteenth century. Prior this, the world view held by physicists was that of the clockwork universe, whereby the entire cosmos was regarded as a mechanism. Each individual particle, however small, was merely a component of a gigantic machine that was slavishly unfolding along some predestined pathway of evolution. According to this image of the universe, nothing genuinely new ever happened, only a rearrangement of the particles. Time, in this context, is reduced to a mere parameter, or as Prigogine puts it: 'God is reduced to an archivist turning the pages of a cosmic history book that has already been written.'

So, in the beginning there was the clockwork universe, a very sterile sort of image. With von Helmholtz's assertion, it became even worse: there was now this degenerative, declining, thermodynamic universe. In both cases, it's very hard to reconcile the world view, either of a clockwork machine-like universe or a degenerating universe, with our image of the world as a coherent and harmonious unity. How can an assemblage of purely passive entities, or worse still, actively destructive irreversible processes, conspire to produce all the elaborate structures and complex organization that is manifested in the world about us? Is the universe in fact dying? What do the observations tell us?

Let me start at the beginning, which, of course, is the origin of the universe ...

Most cosmologists believe that the universe began abruptly in a Big Bang some time between 15 and 20 billion years ago, and that at its origin it was in an essentially featureless state. I think many people have a misconception that the universe was created with everything pretty well intact, with its present structure and present organization somehow imprinted upon it, but that wasn't so. As far as we can tell, in the very early stages it may have been completely devoid of structure, and even of matter altogether. The original state of the universe might have been one of merely empty space containing energy. Everything we see, the matter, the organization of this matter, the structures that surround us, even something as basic as the nuclei of atoms, all these things have appeared since the beginning. Thus, when people talk about the Big Bang as the creation of the universe, it's a bit of a misnomer. It's true that the Big Bang represented the appearance of a rudimentary cosmos, but anything that we would now recognize as the universe has in fact arisen since the beginning. It has emerged in a sequence of steps, phase-by-phase. These steps were very rapid at the beginning and since then they have progressed more slowly. However, it would be quite wrong to suppose that the seeds of the present structure were there at the beginning. No, they have emerged. Hence, this business of referring to the initial event as 'the creation' is, as I say, is a bit of a misnomer.

Complexification

The process of a progressive appearance of structure and organization of the world was referred to at great length by Teilhard de Chardin. He gave it the name 'complexification' and I suppose that is more or less what I mean when I talk about increasing organization and complexity. He believed, and so do I, that pervading the entire cosmos is an inherent property of matter and energy that is defining this progress. As the universe has evolved over billions of years, so new and ever more elaborate structures and systems have arisen. For example, the immense organization represented by a galaxy has formed from a nebulous swirling cloud of gas. The awesome complexity of the biosphere has evidently arisen from a primitive soup of organic molecules. From these and many other examples, it is clear that the universe has never ceased to be creative and Teilhard's insistence that we should no longer speak of cosmology but of 'cosmogenesis' can never have been more apt, though it was certainly unfashionable at the time. The way we see the universe now (in modern cosmology) is as an ongoing creative process, rather than an instant creation event.

Although the progressive character of the universe is really a trend rather than a law, I believe it represents a fundamental principle of nature. One that is yet to be encompassed within our understanding of physics and cosmology. I can express it no better than by repeating Julian Huxley's description of Teilhard's philosophy, which appears in the introduction to *The Phenomenon of Man*. Huxley writes:

> The different branches of science combine to demonstrate that the universe, in its entirety, must be regarded as one gigantic process, the process of becoming, and of attaining new levels of existence and organization.

That emphasis on 'becoming' rather than 'being' is what I wish to draw out later in my talk.

I should point out that similar sentiments to this have been expressed in recent years by a number of physical scientists, perhaps most notably by Ilya Prigogine, the Nobel prize-winning chemist from Brussels, and the philosopher, Karl Popper. Karl Popper sums it up rather nicely, I think, in talking about the 'creativity of nature.' He regards 'understanding this creative power, this creating element' as the greatest riddle of cosmology. Unfortunately, ideas of this sort are frequently misconstrued as somehow standing in violation in the known laws of physics, but this isn't the case. There is no claim, for example, that the second law of thermodynamics is actually invalid. It is only that it's inadequate to encompass this progressive trend that I've been discussing. In fact, biologists long ago discovered how to reconcile the progress of evolutionary change in the biosphere with the operation of the second law. The key to understanding how these things can be reconciled is to distinguish between two sets of things. First, between open and closed systems, and secondly between order and organization.

Systems which seem to go, as Teilhard would put it, against the trend of rising entropy are always found to be systems that are open to their environment. They can export entropy, and transport energy and matter across the interface with their environment. The second law of thermodynamics tells us that, in a closed system, the total degree of disorder must go up. So, taking the universe as a whole, it must go up. However, taking a subsystem within the universe, it is possible for the entropy either to go down or to be maintained at a constant level. Indeed, living organisms are continually exporting entropy into their environment and thus maintaining order. But, that is only part of it, because one must distinguish between *order* and *organization*, and these are often muddled together. Let me give an example: I would regard a crystal as 'order' but without 'organization.' A crystal consists of atoms which are spaced very regularly in a lattice, so it is a very ordered structure. However, it doesn't have very much in the way of organization. By contrast, a living cell is highly organized — an awful lot of co-operative things go on within the cell — but it doesn't look especially ordered, it looks to be a madhouse of activity. It looks very similar to my daughter's bedroom I would say; things thrown all over the place, but she claims she knows where they all belong and what they're all for! So, there is a distinction between order and organization. The second law of thermodynamics refers to *order* and *disorder*, so it is possible to conceive of a situation in which the order of a system decreases (its entropy goes up) but in which the organization also increases.

We have the possibility then of simultaneously having a system in which the order is slowly going down when one takes into account the environment as well, but the organization is going up. The bottom line of this is there is no contradiction between what I am saying about a progressive, evolving cosmos and the imperative of the second law of thermodynamics, which says that the stock of useful energy is running down and the total entropy is going up. However, consistency with the second law doesn't imply that the second law has *explained* this unidirectional growth of organizational complexity. Far from it, those who would believe this, which includes most biologists, fail to make the crucial distinction between order and organization. Let me dwell a bit further on this point, because I think its an important one.

Conventional neo-Darwinism asserts that the engine of evolutionary change is random mutation. I may be preaching to the converted, but chance — the random shuffling of genes — has no directionality. Perhaps it is easier to understand this term 'directionality' with an example. If you are shuffling a pack of cards and you start out in a highly-ordered sequence, you soon produce a disordered sequence. However, starting with a disordered sequence, you have to shuffle for a long time before you get it back into that ordered sequence. During that time of repeated substitution by one random set for another there is no particular directionality in which the card sequences change. Thus, random shuffling doesn't have any inbuilt directionality. So, there is a fundamental problem about the arrow of time in biology.

Some biologists have attempted to evade this problem by declaring that there is no evolutionary advance. They say that there is no biological meaning in proposing that human beings, for example, are in any sense better than bacteria. They believe that this is a chauvinistic view of the world. But, whether or not human beings are somehow better than bacteria, they are certainly more complex and more highly-organized. So, the unidirectional increase of complexity and organization defines an objective arrow of time.

Well, let me summarize then: We have these two cosmic arrows of time, one which seems to lead to degeneration and disorder, and the other which seems to lead to progress and increasing organization. In recent years, scientists have come to recognize that matter and energy possess an innate ability to self-organize and that this ability is manifested in all branches of science, not just biology. I think that this is really very significant. We recognize what we might call the 'miracle of life' — the ability of living organisms to increase their complexity and organization and that baffles us all. However, these self-organizing processes are also to be found in systems which are regarded as inanimate, and the study of this self-organization is a multi-disciplinary subject with many practical applications.

Self-organization

To focus attention on what I mean by 'self-organization,' I'll give you one or two examples. There is one very homely example, literally 'homely' in the sense that you can do it in your kitchen. Consider liquid in a vessel which is heated from below, something that happens in every kitchen every day. Initially the fluid is in a uniform featureless state. When the temperature difference between the bottom and the top of the fluid reaches a certain critical value, the system flips abruptly into a new state: it starts to convect. The convention isn't just any old movement of the molecules, but an exquisitely organized affair. In fact, if this experiment is done carefully in the laboratory, the convective motion establishes itself into a very distinctive pattern of regular hexagonal cells.

The most remarkable feature about this orderly state is that the cells themselves are very many orders of magnitude larger than the distance between the molecules and the range of the forces that act between them. Each molecule merely pushes or pulls its neighbour. Yet, somehow, the total assemblage of molecules in the pan cooperate to produce this globally-coherent pattern of activity. No external agency says to the water molecules 'you must move this way and cooperate with that molecule over there,' and so on. The whole thing happens spontaneously. The long-range order establishes itself; Prigogine describes this as 'order arising out of chaos.' There is no contradiction with the letter of the second law of thermodynamics, although it contradicts the spirit of the law to go from uniformity to something elaborate with structure. A key point is that the self-organization of the liquid into hexagonal convection cells happens when the system is driven a long way from equilibrium.

Another well known example of spontaneous self-organization is the laser. In an ordinary lamp, light is emitted in little bursts, each totally unrelated to all the other little bursts. So, the light that comes out of a lamp is all jumbled up, with all the waves incoherently superimposed. However, in a laser things are very different. What happens there is that the system is once again driven far away from equilibrium. (A laser is a highly non-equilibrium system, whereas an ordinary lamp is close to thermodynamic equilibrium). In the laser, all of the individual little wavelets conspire; billions of atoms cooperate to emit their light exactly in phase — it is like soldiers marching together in step — to produce a coherent waveform. Laser light is a very pure, exact sinewave that can stretch for hundreds of thousands of miles .

So, here we have two relatively familiar examples of how really quite mundane physical systems, in one case fluid heated from below and the other case a collection of atoms which are just pumped or excited, when driven away from equilibrium spontaneously, organize themselves into a state of higher organization and complexity.

Examples of self-organization can be found not only in biology and physics, but in all the sciences. Prigogine has drawn attention to a number of intriguing non-equilibrium chemical reactions where large scale co-operative behaviour is displayed. For example, there are so-called chemical clocks where mixtures of substances spontaneously settle into a rhythmic pattern. You can arrange it so that at one moment the liquid looks blue, then the next moment it looks red, then it looks blue, then it looks red, and so on. It ticks away like this in a very precise manner. Once again you have the mystery that, right across the body of this fluid (which may be many centimetres across, much greater than the separation between the molecules) there is this co-operative behaviour taking place. It is as though each molecule thinks: 'What are the others doing? I must keep in step.'

Even in astronomy, we can recognize systems in which there has been some attempt at co-operative behaviour, a settling down into preferred modes or patterns. A familiar example is the Great Red Spot of Jupiter. This is an example which Teilhard de Chardin would have liked, because it is a process structure. The Red Spot is not a lump of something floating around in the atmosphere of Jupiter, it is in fact a region in which gases are continually flowing in out. It has a throughput of matter all the time, and yet it retains a coherent identity. Computer models of the Red Spot suggest that if something were to cause it to break up, it would rapidly reassemble itself. So, it's got a certain robustness to it. The Red Spot is an example of many stable structures, or quasi-structures, which can occur in fluid systems.

If you look closer at the detail surrounding the Red Spot and other regions in the surface of Jupiter, you see examples of exquisitely fine detail. Again, let me stress that this is not a solid surface we're looking at, it's a fluid, but the fluid has arranged itself into vortices, whorls and all sorts of intricate forms. It is reminiscent of the pattern of flow of a stream, with eddies and swirls. It leaves you wondering about the creating ability of the nature to bring forth so many different coherent shapes from featureless uniformity.

Another very familiar astronomical example of self-organization is the rings of Saturn. The rings are not some solid plate, but are made of myriads of small particles orbiting Saturn together. They have organized themselves into an incredibly intricate structure of rings, and rings within rings, and rings within rings within rings. This is a very large degree of internal structure. No-one has made the rings of Saturn. It has established itself spontaneously.

Of course, it is in biology where the most striking examples of self-organization are found, most notably in the ability of the embryo to develop from a single cell via an exquisitely orchestrated sequence of steps. However, it would be wrong to suppose that it is only in biology that we see such sorts of self-organizing changes occurring.

What do these systems have in common? We live, evidently, in a universe which is replete with self-organizing system. The universe as a whole, I've been arguing, has a progressive self-organizing character. So the question arises: can we see any sort of general laws or principles at work? Certainly there are a number of conditions that seem to be necessary before a system displays self-organizing behaviour. First, they are invariably open systems, with an exchange of material and energy with their surroundings. They are not static structures, but processes. I gave the example of the Great Red Spot of Jupiter as a process structure with material and energy flowing in and out all the time. A living cell would be another example. Even a cyclone on the surface of the earth can be regarded as a primitive process structure. The point is that these are not static, closed systems, but they are open to their environment. However, in spite of the throughput of matter and energy, they nevertheless retain a certain discrete identity. We can recognize the Red Spot as it comes round each time, even though the material within it may not be the same on every occasion. Similarly, we can recognize a person from year to year, even though the atoms of the human body are continually being replenished.

Another feature of self-organizing systems is that they tend to be what is called 'dissipative,' that is to say, any disturbances which act upon them are rapidly dissipated away. This can be traced to the fact that they are far from thermodynamic equilibrium, thus acquire a robustness and stability. But, perhaps the most crucial feature of self-organizing systems is that they are what mathematicians call 'non-linear.'

Understanding the weather

For three hundred years, scientists have been mainly preoccupied with so-called 'linear systems.' Roughly speaking, linear systems are those in which cause and effect are associated by some simple proportionality. Let me give an example: a piece of elastic. If you stretch it, then the amount of stretch that you get is in simple proportion to the amount of force you apply. It's a linear system, so cause and effect are linearly related. If you plot the cause against effect on a graph, you get a straight line, hence the word 'linear.'

Many systems are approximately linear, and they are, mathematically speaking, relatively easy to describe — which is why so much progress has been made in understanding them. It is often said that a linear system is nothing more than the sum of its parts; a conjunction of components in peaceful co-existence. Complexity arises in linear systems merely as a result of superposition. If you put a lot of simple things together, you can get a complicated thing. That is, of course, a rather trivial form of complexity. Radio waves are a very good example. A radio wave from a transmitter, which is a familiar linear system, can have a very complicated form when it's travelling from the radio station to the receiver. When the wave is picked up at the receiver, it's a complicated form — all the different frequencies superimposed— are filtered out by the radio, broken down or analysed into a simple set of components. You can put the original waveform back together again if you like; linear systems are reversible. The point about linear systems is that they can be taken apart and reassembled; the individual components don't get in each other's way. Linear systems are simple to study, precisely because they can be chopped up and analysed in this way without destroying their essential nature. The philosophy of reductionism therefore works very well for linear systems.

By contrast, a non-linear system is more than the sum of its parts. There is, to quote Teilhard de Chardin, a 'creative value of synthesis.' Synthesis in a linear system is mere superposition — two or more things put together without interaction. However, in a non-linear system, a and b together make more than $a+b$. The conjunction of two components in a non-linear system can lead to novel and very often unforeseen effects. Different parts of a non-linear system often interact to produce something totally new. Non-linearity arises for a variety of reasons often as a result of feedback mechanisms or from breakdown of the simple proportionality in the action of forces. In reality, all systems are non-linear, it's just that many are to a good approximation linear, so that the methods of linear analysis work very well for a wide range of systems.

In a non-linear system, complexity is much more than mere complication, more than a lot of things jumbled together. Even very simple non-linear systems can display remarkably complex behaviour. A classic example is provided by something that you would think ought to be very simple: a pendulum. The expression 'as regular as clockwork' epitomizes the notion of mechanical simplicity and dependability; remember the clockwork universe? Well, you would think a pendulum ought to be a very good example of part of the clockwork universe in action. That is certainly true of an ordinary pendulum that just swings backwards and forwards in a plane. It doesn't do anything exciting, it keeps regular time. However, if you allow the pendulum to swing, not just in one plane but in two like a ball on a string, then things can become quite complicated. It turns out that for a large range of driving frequencies, all that happens is that after a time, the pendulum settles down to a regular pattern of motion, describing a sort of ellipse. You could go away and you can come back the next day and it would still be chugging its way round. It's an extremely predictable, regular type of motion. However, if you turn the frequency up a bit, and drive the pendulum at just above its natural frequency, something extraordinary happens. Instead of settling down to a regular pattern of motion, the pendulum goes haywire. It sometimes goes this way, and sometimes goes that way. Sometimes it moves anti-clockwise and sometimes clockwise. If you look at it, say, for five of six swings and place a bet as to what it will be doing in one minute's time, it can be shown that the betting odds are no better than evens. The pendulum is completely random in its behaviour. This is an example of a so-called 'chaotic system.' The behaviour is completely random and unpredictable, even though it is a deterministic mechanical system of a type we might imagine can be predicted in great detail.

Chaotic systems are now the subject of much study. It's a very fashionable subject and people have found chaotic behaviour in a huge variety of physical systems. One example is the weather. Now maybe people will be easily persuaded that the weather is chaotic. This is, of course, bad news for weather forecasters. Using a very simple mathematical model of the weather, it is indeed found to be chaotic. One day's weather forecast is roughly like one swing of the pendulum — pretty predictable. But the weather five or six days hence is essentially unpredictable, at least in its finer details.

The unpredictability here is not just a matter of our ignorance of the forces at work. It's not that the weather or the pendulum is intrinsically complicated. In a chaotic system, the behaviour is exquisitely sensitized to the initial conditions. This means that, when the system is starting off, if your input information is only very slightly different from the truth, your prediction very rapidly becomes worthless. Two identical chaotic systems with imperceptibly different starting conditions soon diverge in their behaviour. The divergence is an exponentially increasing process, so that after a very short period of time, you've lost all predictive power. If you increase your information gathering ability, you encounter a law of diminishing returns: if you get ten times as much information about the initial state in order to be twice as accurate on your forecast, then a hundred times information might make you only four times as accurate and so on; you very rapidly run out of computing power and money. There's no point in having larger computers to forecast the weather, you're flogging a dead horse.

I've been talking about non-linear systems which are driven away from equilibrium, that can spontaneously and abruptly leap into new, more highly organized states. Or they may degenerate into chaos. In a sense, matter and energy have some sort of free will, an ability to explore alternatives which could have not been foreseen. In a linear system, one can foresee what they are going to do, but in a non-linear system such as the pendulum which is free to swing in two directions one simply doesn't know what the system is going to do.

Traditionally, scientists have tended to de-emphasise complex systems, treating them as a sort of annoying aberration. This reflects the fact that non-linear systems are hard to study. However, by focusing on simple linear systems at the expense of non-linear ones, science has developed a strongly reductionistic bias. It used to be argued that complex systems are, in principle, always explicable in terms of their components — you just have to chop them up and look at the individual components and then you will understand them as a whole. This purely methodological reductionism became so ingrained that it translated itself into a genuine belief: The only truly fundamental statements about the world are those that refer to the bottom level entities in nature, the elementary particles and fields of physics. Indeed, the very word 'elementary' here implies that the primary phenomena of nature are those pertaining to individual irreducible objects alone. Collective phenomena, such as self-organization, life processes, turbulence in fluids, super-conductivity, conscious awareness — all these things have come to be regarded as purely secondary and to be reducible to, or derivable from, the primary phenomena with no additional principles being necessary. This is the illusion that Teilhard referred to.

Unfortunately, the entire vocabulary of science has become rooted in this reductionistic philosophy. The concept of causation for example, is almost invariably applied in the 'upward' sense, that is changes in complex macroscopic systems are regarded as caused solely by the activity of their constituent atoms. Conversely, explanation is generally considered to apply 'downwards.' For example, psychology and sociology are said to be explained by, or reducible to, biology. Biology, in turn, is explained by chemistry and thence by physics until ultimately all of physical reality is founded solely upon the activity of elementary particles and fields. According to this philosophy, the only real point of contact between the sciences is at the bottom level and the only true laws are the laws of particle physics. It's a position that is eloquently summarized by Leon Lederman, director of Fermilab, a particle accelerator laboratory in America: 'We hope to explain the entire universe in a single simple formula that you can wear on your T-shirt.' That is the great hope of the reductionist programme, that everything will be reduced to a simple formula.

Well, many distinguished scientists, including a number of famous Nobel prize winners, have rejected the reductionist picture of the world, although they have tended to be dismissed as sentimentalists or vitalists. I believe, however, that a sea-change is taking place in the scientific community due to the rapid progress being made in understanding non-linear systems, complexity and self organization. Across a whole range of disciplines, a new paradigm is emerging. This transformation is due in large part, I might say, to the advent of fast electronic computers, which have made possible for the first time the study of complex non-linear systems in great generality. As a result of this explosion of knowledge, a curious inversion of outlook is taking place. Complexity is now seen as the norm rather than as an aberration; most physical systems, as I have remarked, are in fact complex non-linear open systems. The simple closed linear systems of traditional science are now recognized as extreme idealizations belonging to a very special class. The strong emphasis accorded to such systems in the past can be regarded simply as a sort of selection effect. When scientists had available only primitive techniques and studied only simple systems.

The essentially holistic nature of complex phenomena is well illustrated by the burgeoning study of neural networks, or networks theory generally. A network consists of an array of what are called 'nodes' connected together by wires, or lines, or something. This could be a model of the brain, where the nodes are neurons linked by axons and dendrites, or it could be a model of a magnetic material, where the nodes are atoms that can be oriented in one direction or another in the lines of a magnetic field. A wide range of systems can be modelled as a network like this. In a simple idealization each individual node can be in one of two states: 'on' or 'off'; for example a neuron firing or not firing. The study of networks consists of taking an initial state — that is, specifying which of the nodes are on and which are off — and specifying how they interact via the connection lines. You can then evolve the mathematical model deterministically to see what happens. The result is a great surprise. You might think that if you just start out with a random state, then you couldn't expect anything very interesting to occur. However, some networks show remarkable self-organizing qualities. Stable patterns of activity swirl around the network and settle into organized forms. Networks provide classic examples of holistic systems. If you remove a linking line, you don't destroy the whole pattern, the pattern has a robustness and stability. Remarkably, the pattern itself is a quality which lives on the network as a whole, it's not associated with a state of any given node and if you knock out a particular node it doesn't terribly much interrupt the pattern of behaviour. That is why our brains can function even though many individual neurons die. When one is studying these networks, one is studying the collective organization of properties of the network as a whole, and not the state of any individual node or connection.

In describing the behaviour of a network, it is clearly meaningless to talk about reducing the network down to the behaviour of individual nodes, because the very definition of a network is the collective interconnection of things. So, here we have a good example of a burgeoning study, the study of networks, which is very much part of the artificial intelligence programme, designing smart computers that mimic the operation of the brain. It's a very practical sort of thing , and yet it well illustrates the new way of thinking about certain phenomena in terms of their collective and organizational properties.

How long is a piece of string?

There's another fashionable example of complexity which I might mention very briefly, known as fractals. They were first discussed by Lewis Fry Richardson, the uncle of Sir Ralph Richardson, many years ago. He was interested in the thorny problem of the length of the coast of Britain. If you think about it, you'll soon see that the answer is not very well defined. Ask what's the distance between Southampton and

Land's End? If you look at an atlas you can get a rough idea of what that distance might be, not forgetting to go in and out of the little coastline wiggles. However, if you now get an Ordnance Survey map you will see that there are many more wiggles than there are on an atlas. If you've got an even larger scale map you'd see that there would be still more wiggles. The finer and finer the detail you examine the system in, the more and more wiggly it becomes and the longer the total distance becomes. If you entered that contest in last week's *Observer* to win a holiday to Mauritius, you had to guess the length of the coastline of Mauritius. The correct answer is actually infinity. (I hope the judges know that, because I'm looking forward to my holiday in Mauritius!) This insight of Lewis Richardson was that the length of the coastline is not well defined, because you're trying to measure the length of something that is infinitely irregular.

This has now all been put on a sound mathematical footing by Benoit Mandelbrot of IBM, who invented the term 'fractals' for a structure which is irreducibly complex in this manner. If we take any given section of that same coastline and look at it in greater detail, it has the same degree of wiggliness as the whole. In other words, any part of it in some sense 'contains' the whole. One can make this mathematically very precise and display mathematical functions that have this exact self-scaling property. There are very many structures besides coastlines that have been identified in nature which are quite clearly fractals.

The points about fractals is that they are primitively and irreducibly irregular. In the old way of looking at the world, people would always suppose that, when you had something complicated, the thing to do would be to look at it in ever finer detail until you eventually get to its simplest components. However, a coastline, or any other structure that has this infinite irregularity, shows the folly of that view. Many structures that occur in nature are irreducibly irregular, and instead of trying to model them very imperfectly by approximating them to regular forms, one should accept irregularity and complexity as a basic component in nature. So we see once again how anti-reductionist or holistic thinking is penetrating science. People now are identifying fractals in a wide range of different sciences and recognizing that they are primitive in their own right, they are just not reducible to smooth or simple structures.

Organizational principles

I have been talking a lot about complexity and organization as found in the different branches of science and the message that I want to put across is that, although we find many examples of these things in different sciences, there are certain universal principles that appear to weave them together. Regularities are found, implying the existence of organizational principles of a very general nature. I mentioned, for example, that the network concept could apply either to the brain or to magnetic materials, or to evolutionary strategies in biology. There is a large range of things that, basically, the same network can model — the organizational properties of the network can be manifested in very different fields. There is now a real hope, I believe, that the age-old conceptual conflict between physics and biology might be nearing an end. Biological processes like morphogenesis and evolutionary development, which seem miraculous at first sight to the physicist, can be viewed as but one (albeit distinctive) example of general organizational principles — principles, as I pointed out, that can be found in inanimate systems too. I believe that orthodox neo-Darwinism, whilst correctly identifying the basic mechanisms of evolutionary change, nevertheless fails to capture the essential organizing element that generates the progressive arrow of time I referred to. However, if we combine neo-Darwinism with the new mathematical principles that have been discovered in inanimate systems, then I think we will have a satisfactory explanation of what we might otherwise perceive as a mystery.

Now, just as there seem to be general organizational principles of order, so there are general principles of disorder. It has been discovered, for example by Mitchell Feigenbaum, that the onset of disorder in a very wide class of systems displays certain universal features. A system doesn't simply become chaotic, its approach to chaos occurs in a sequence of mathematically precise steps, in the pattern of a fractal in fact. The scaling behaviour of these steps is characterized by two numbers, which for the benefit of those who want to copy them down are 3.5699 (to the first four decimal places) and 4.6692. These are just two unremarkable numbers and you might think they are not especially exciting. They are called 'Feigenbaum's numbers,' of course. They are universal numbers in that they occur wherever we see chaos arising, at least by one particular route, whether it's in some turbulent fluid, a pendulum, a population of fish, or whatever. So, even in chaos there is regularity. I would argue that Feigenbaum's numbers are fundamental constants of the world and yet they can't be reduced in any way to the other constants of physics. They are numbers that refer to the collective behaviour of complex systems, but they're constants that can't be reduced to the behaviour of the components of those systems. The remarkable appearance of these general regularities in complex organized systems provides what one would think of as a high-level bridge between widely different physical phenomena and different branches of science. It's not merely at the level of elementary particles, the formula on your T-shirt, that we connect everything together. We find these interconnecting regularities at a much higher level as well, and I see no particular reason to dignify the former connection, but not the latter, with the label 'fundamental.'

I hypothesise, then, the existence of fundamental principles that govern organization and complexity, but that doesn't mean that I'm denying the validity of the lower levels of the laws of physics. It's possible to have consistency between the higher level principles and the laws of physics. Let me give a specific example of how there can be such consistency. When I play Space Invaders on my computer, the shapes on the screen move about according to certain rules. That's the whole idea of the game. Now the laws, if I might call them that — these dynamical rules that affect the shapes on the screen — are programmed into the computer. It would be absurd to say that, just because the computer operates according to the principles of physics that the laws of the space invaders' shapes can be reduced to the laws of physics, that is, the laws that govern the motion of electrons. But this is not to say that the shapes on the screen are in some way violating the laws that govern the motion of the electrons. It's just that the laws governing the motion of the electrons are inadequate to explain the motion of the shapes on the screen. The secret of the computing machine lies in the hard-wired configuration of its circuitry and in the software that's programmed into it. Boiled down to its physical essentials, these features amount to a pattern of constraints on the physical processes that occur in the hardware. These constraints aren't, of course, encompassed within the laws of physics themselves. They can't be reduced to or derived from, these laws. They must be imposed in addition to the laws by considering the design details of the specific system concerned. It doesn't make them incompatible with those laws, nor is there any incompatibility between the space invader dynamics and electronic dynamics, but nevertheless the one can't be reduced to the other.

There is a rather nice quote from Marvin Minsky, who is one of the world's experts in artificial intelligence, in which he addresses this issue. Minsky, in his recent book says:

> Many scientists look on chemistry and physics as ideal models of what psychology should be like. After all, the atoms in the brain are subject to the all-inclusive laws which govern every other form of matter. Then, can we also explain what our brains actually do entirely in terms of those same basic principles? The answer is 'no' simply because, even if we understand how each of our billions of brain cells work separately, this would not tell us how the brain works as an agency. The laws of thought depend not only on the properties of those brain cells, but also on how they are connected. These connections are established not by the basic general laws of physics, but the particular arrangements of the millions of bits of information in our inherited genes. To be sure, general laws apply to everything, but for that very reason they can rarely explain anything in particular.

I think that quote nicely brings out how one can look for additional general principles, principles that connect things at a higher level, without violating or getting in the way of the lower level laws.

Being and becoming: two concepts of God

Rupert Sheldrake will likely refer to the tension that apparently exists between two contradictory images of the world, which we might call the 'changeless' and the 'changing,' or 'being' and 'becoming' as it is fashionably described. I have been talking about a picture of the universe which emphasises the progressive evolutionary character, the 'becoming' aspect. However, we somehow have to square this picture with the other, a timeless, view of the world. This clash of the temporal and atemporal is very deep. It appears in both science and religion, and it's worth dwelling upon briefly here.

First of all, let me say something about religion. I'd like to suggest two very different images of God which are often conflated: God the Creator and God the Father. God the Creator is the remote powerful agency who can do things like create laws and bring universes into being, whereas God the Father is the personal inner God, the God of prayer and love. The key distinction in my opinion between these two images has to do with their relationship with time. God the Creator must, by his very nature, transcend both space and time. The laws of physics, if they're anything at all, are timeless abstract relationships. Let me stress that if these laws are to provide an explanation for how the universe came into being, as I believe they do, and if time is part of that universe, which it is, then they must exist timelessly in some abstract realm. So, perhaps calling the God that is responsible for those laws God the Creator is a bad term, as 'creation' implies an act within time. Perhaps God the Sustainer is a better way of putting it. Now, I believe all the world's great religions have arisen out of an attempt to reconcile the apparently irreconcilable, namely the changing and the changeless. Let me justify this by appealing to our most primitive experience of all, which is our experience of ourselves.

The concept of the self is, of course, a notoriously slippery one. What precisely do I mean by 'me'? Clearly the self would have no significance if it wasn't in some sense preserved in time. If I woke up each day a different person, the label 'I' really wouldn't mean very much. There must be some sort of unchanging quality running through it. It's clear that memory must play some a part in this, because, if I couldn't remember from one moment to the next what I was doing, then the notion of my personal identity would be meaningless. Something must remain constant. On the other hand, my experience of myself involves thinking and observing, and these are clearly temporal activities. Every day, I have new experiences of some sort and, in this sense, every day I change a little bit. You might say I'm not the same person now as I was at the age of five, and yet the fact that I used the word 'I' in this sentence reveals the paradox. How can 'I' be a changed person if I am the person to whom the change is occurring? Life would be pointless if there were no change, no progress, yet life would be meaningless without a sense of personal identity — a 'me' for these changes to happen to. The most severe change for us all is death. Here we see the same tensions recurring, but with a vengeance, because many people fervently wish their personal identity to survive death. Yet survive for what?

If it's to join some supernatural realm beyond time, we lose those very things — thoughts, experiences and so on — that give us a sense of continuing personal identity, for these are all temporal things. On the other hand, it's possible to imagine passing on to some realm *within* time, such as through reincarnation within this universe. But, if I carry over no memory of my past life into my new one, what right have I got to say that it is *me* that is being reincarnated rather than somebody else? Well, perhaps there's some supernatural world with its own version of time as well as presumably space and people, but if that's the case, can one cope with the prospect of an *infinite* duration of changing experiences? Are there, in fact, an infinite number of things to experience?

These temporal tensions, if we can call them that, spilled over into theology a long time ago. The most striking manifestation concerns the issue of creation. In ancient Greece, for example, one school of thought, led by Aristotle, maintained that there had been no creation, that the universe had endured for an infinite time. However, all cultures also have their creation myths in which some sort of super-natural agency creates the present harmonious order out of primeval chaos. In the Christian era, this issue resurfaced with the question: 'What was God doing before he created the universe?' Given that he had waited an infinite time to act, why did he choose to create the universe at that particular moment rather than at some other? A deep problem. Well, Augustine, who thought very deeply about the nature of time, had an answer to this. He proposed that God didn't wait an eternity before he created the cosmos. Instead, he created time as part of the universe. To use Augustine's own words: 'The world was created with time and not in time.' That was remarkably prophetic of the modern scientific view of the creation, which involves the creation of space and time as well as matter. The idea of God sitting there, within time, for an infinite duration and then creating the universe at an arbitrary moment just isn't on. So this type of Creator God, or God the Sustainer, must exist outside of space and time altogether. However, Augustine's solution only served to create another problem, because by placing God outside of time completely, there is a conflict with the other image of God, which I have called God the Father.

People who believe in a personal God want him to communicate with them by answering prayers, watching over them, guiding them, getting angry with them, forgiving them, judging them and so on. These are all indubitably temporal activities. The difficulty is particularly acute in those religions that demand some direct divine intervention in the affairs of the world, because, of course, this means that God chooses to act at some particular moment in time. So, on the one hand one would like some sort of image of a God the Creator, a timeless being of utter consistency and dependability, an assured Absolute that underpins all of existence. On the other hand, one may also like the idea of a God who can in some sense change his mind; to answer prayer, for example, or avert disasters.

All religions have had to confront this paradox of the changing yet changeless Deity, and they have done this by employing all sorts of abstract notions and subtle metaphors, or by coming down on one side or the other. I began my address by quoting from Teilhard de Chardin. As has been the theme of my talk, Teilhard has emphasised the progressive, evolutionary nature of the cosmos — the flowing, moving, unfolding nature of time. Nevertheless, he proposes that the culmination of this progression is a mysterious Omega, which itself lies outside of time. In fact, Teilhard says Omega must be supremely present. So, in some way, Omega both crowns a temporal series and timelessly represents the closed nature of that series. De Chardin somehow manages to have something which is both within time and outside of time. It's a very difficult concept. Many Eastern religions also grapple with this issue of the changing and the changeless by having a cyclic view of time. Things are locally progressing but, looked at on a longer timescale, they are cyclic. In mysticism, both East and West appear to confront this issue of the temporal and atemporal in conjunction by attempting to achieve a timeless mental state within time, thus resolving the paradoxical conjunction of the temporal and atemporal.

Interestingly, science has undergone a parallel debate about the question of the changeless and the changing, of eternity and the flux of time. I've tried to draw out some of the elements of that debate this evening by showing that, historically, there have really been three scientific paradigms:. The first was the paradigm of the clockwork universe, in which time is irrelevant, and God was the mere archivist, where everything was laid down once-and-for-all and every state of the universe followed deterministically from every other state. The second was the thermodynamic view of the universe, that takes into account change and flux, but it is a degenerative type of change. Finally, we have the more hopeful progressive universe, the universe of self-organization and increasing complexity. However, the tensions between the two views of time crop up in all of these world views.

Well, let me draw this to a close; on the subject of time. I want to argue that the tensions that exist between the two views of time aren't just a matter of something to be resolved by further enquiry. I don't think that, as we get to know more about the world, we are going to say that one view of time is right and the other view is wrong. I think that we have to live with these two apparently contradictory images of time, and that it's an essential part of our understanding of the world and ourselves. We have to see things in these two apparently contradictory ways. Now, it's fashionable to declare that the advance of science has pointed out humanity's insignificance, even irrelevance, in the immense universe. An awareness of the size of the physical universe and the vulnerability of our tiny planet have given rise to the idea that humans, in particular, and life in general, is merely some sort of quirk of nature, an incidental feature in the workings of a complex universe, a totally meaningless accident. Well, the discovery of the creative cosmos, which I have been outlining here, has changed all that. The self-organizing capabilities inherent in matter show that, far from the emergence of life being a mere accident, it's an integral stage in the continuing process of matter and energy achieving states of greater and greater organizational complexity. The fact that we live in a creative and progressive cosmos is now at last scientifically respectable. This inevitably raises the question of whether there's a meaning to it all. Is the universe unfolding according to some plan? Is there a cosmic blueprint, as I like to put it? Well, the idea of the universe manifesting some sort of design is a very old one. It seems to me that the most compelling evidence for design is that the laws of physics are constructed in such a way that they permit the self-organizing qualities to exist. Furthermore, that these self-organizing qualities not only give rise to interesting structures of the sort that I have been describing, but they give rise to conscious individuals who can then try and work out these laws and ponder what it all means.

So what does it all mean? Well, science of course can't answer questions about meaning, only about mechanisms. There will always be some scientists who'll see in the creative cosmos nothing but a pointless charade. Others, however, will find these new developments deeply inspiring and it will confirm their belief that there is a meaning behind existence. In the end, it has to remain a matter of personal belief. I do think that whatever one's own personal beliefs, the discovery that human beings are an integral part of an ongoing complexification process (as Teilhard de Chardin chose to call it) bestows on us at least a certain dignity, a dignity we would lack if we were merely an incidental accident amid the vastness of the cosmos.

The Experience of Unity:

At the Origin of Religion, Philosophy, and Psychology, and at the Ultimate Horizon of Physics, Cosmology, and Consciousness

KURT DRESSLER

Professor Kurt Dressler was Vice-Rector for the Doctorate (Dean of Graduate Studies) of the Swiss Federal Institute of Technology (ETH Zurich) for seventeen years. Born in 1919, he received his Ph.D. in experimental physics at Basel University. After twelve years of research and teaching in atomic, molecular, optical and astrophysics at N.R.C. of Canada in Ottawa, at N.B.S. in Washington, D.C., at Mt. Wilson Observatory in California, and at Princeton University, where he participated in the design of a space telescope, he was appointed Professor of Molecular Spectroscopy at the ETH in 1967. Currently he is lecturing and writing on unification of science and religion, on the search for unity in philosophy, unification physics, 'big-bang' cosmology, consciousness studies, and religion, and on the significance of the realization of unity consciousness in experiences of mystical union. This lecture was delivered at the 1997 Mystics and Scientists conference.

Before I attempt to talk on the experience of unity, I need to prepare my mood, and maybe yours, for such an undertaking. I shall do that by reading a passage from a book that for me ranks as a sort of a 'modern holy scripture.' The particular passage which I'm going to read has been called 'The Forgotten Song.' It is a text which speaks to the heart as well as to the intellect; I think it will speak to us as mystics and as scientists. At first hearing it appears to be just another poetic spiritual text, but closer examination will show us that it simultaneously is of significant psychological, philosophical, and even physical and astronomical relevance. I have chosen to read this text because of its relevance to the title of this talk.

Listen now, and try to think if you remember what we will speak of now.

Listen, — perhaps you catch a hint of an ancient state not quite forgotten; dim, perhaps, and yet not altogether unfamiliar, like a song whose name is long forgotten, and the circumstances in which you heard completely unremembered.

Not the whole song has stayed with you, but just a little wisp of melody, attached not to a person or a place or anything particular.

But you remember, from just this little part, how lovely was the song, how wonderful the setting where you heard it, and how you loved those who were there and listened with you.

The notes are nothing. Yet you have kept them with you, not for themselves, but as a soft reminder of what would make you weep if you remembered how dear it was to you.

You could remember, yet you are afraid, believing you would lose the world you learned since then.

And yet you know that nothing in the world you learned is half so dear as this.

Listen, and see if you remember an ancient song you knew so long ago and held more dear than any melody you taught yourself to cherish since.

Beyond the body, beyond the sun and stars,
past everything you see and yet somehow familiar,
is an arc of golden light that stretches as you look
into a great and shining circle.
And all the circle fills with light before your eyes.
The edges of the circle disappear,
and what is in it is no longer contained at all.
The light expands and covers everything,
extending to infinity forever shining
and with no break or limit anywhere.
Within it everything is joined in perfect continuity.

Nor is it possible to imagine that anything could be outside,
for there is nowhere that this light is not.

This is the vision of the Child of God, whom you know well.
Here is the sight of him who knows his Creator.
Here is the memory of what you are; a part of this,
with all of it within, and joined to all as surely as all is joined
     in you.
Accept the vision that can show you this, and not the body.
You know the ancient song, and know it well.
Nothing will ever be as dear to you as is this ancient hymn of
     love
the Child of God sings to his Creator still.
    What is a miracle but this remembering?
And who is there in whom this memory lies not?
The light in one awakens it in all.
And when you see it in your brother, you are remembering for
     everyone.[1]

I shall refer to the various parts of this so-called 'Forgotten Song' as
we go along. I have structured the talk into seven parts that correspond
to seven aspects of the title:

1. The experience of unity:
   The vision of interconnectedness and the experience of mystical
   union;
2. at the origin of religion:
   The union of Self with All, experienced as union with God, at the
   origin of religion;
3. of philosophy:
   The distinction between maya *and truth at the origin of*
   philosophy;
4. of psychology:
   The psychological evolution from unconscious to conscious
   unity;
5. and at the ultimate horizon of physics:
   The idea of unity in physics —
       5.1 Physical reality is an unbroken whole;
       5.2 Unification of space, time, energies and forces;

6. of cosmology:
  The concept of unity at the ultimate cosmological horizon;
7. and of consciousness:
  The unification of physical reality and consciousness.

1. The vision of interconnectedness and the experience of mystical union

'The Forgotten Song' never fails to touch me emotionally as well as intellectually, intuitively as well as rationally. It always reminds me that there is no separation between self and other, between myself and what I call the world. 'Everything is joined in perfect continuity;' my body, my consciousness, the world: all are 'joined as surely as all is joined in me.' The all-encompassing interconnectedness that is formulated here is a well known characteristic of the experience of mystical union. In mystical union all multiplicity 'returns to its origin, which is One without any duality, and which comprises the multiplicity.' Entering into mystical union one 'returns from multiplicity and separation into the original unity.'[2] The experience is essentially indescribable because who experiences it has become a unity. 'The consciousness of the One comes not by communicable knowledge, but by an actual Presence superior to any knowing.'[3]

2. The union of Self with All as union with God at the origin of religion

The experience of mystical union is at the origin of religion because the vision of interconnectedness and the union of Self with All is most often recognized as union with God. Martin Buber has written that:

> there is almost no mystic who would not interpret her or his experiencing the true undivided self as experiencing God. In mystical union Man receives a revelation of freedom and of undifferentiated presence: the experience of self. But he doesn't dare to lay it on his own poor self, being unaware that it supports the world-self. So he lays it on to God. The soul who is given the grace of experiencing unity is no longer caught in the multitude of distracting perceptions of self and world, but she experiences the unity of self and world.

> Normally the soul seems to lay on to God the explanation of
> everything she cannot understand of the world. But after the
> experience of mystical union she lays on to God the
> explanation of what she can least understand in herself: the
> experience of unity. This now becomes God's highest gift.[4]

A friend of mine says that 'the deepest reality of humanity is
experienced in mystical union. It reveals, at the very deepest level, that
we are divine, we are all brothers and sisters.'[5]

In his book, he quotes Kolakowski: 'The experience of mystical
union is a rare phenomenon, but it makes up the core of religious life.
It is the only direct and absolutely irresistible experience of the Eternal
and Infinite as such. It is at the source of the great religions and it is
decisive in keeping mankind's religious legacy alive.'[6]

And Heschel: 'The essence of religious thinking does not lie in
entertaining a concept of God but in the ability to articulate a memory
of moments of illumination by His presence.'[7]

I now turn to the observation that the 'experience of unity' is not
only 'at the origin of religion' but also 'at the origin of philosophy.'

3.   The distinction between maya *and reality at the origin of philosophy*

I claim that the concept of unity is the answer to the fundamental
philosophic search for the origin, essence, and end of our puzzling
existence in a multifaceted world.

Two and a half millennia ago the pre-socratic Greek philosopher
Parmenides drew radical philosophic consequences from his own
mystical experience. He stated flatly that:

> philosophic truth is in stark contrast to the way we ordinarily
> perceive the world. For, firstly, conventional wisdom imagines
> that the single thing in its individuality bears true reality and
> significance, disregarding the Whole to which the single thing
> entirely owes its existence. And secondly, conventional
> wisdom imagines the world as a conflict between opposites,
> forgetting that in each conflict there is a Unity on whose
> ground alone the opposites can arise. True Being — unlike
> finite being — is not split into single things, but it is One. In

It all in all is interconnected. It knows neither contradiction nor fight, but it manifests wholeness, inseparability, sameness with itself. It is characterized by neither corruptibility nor incessant shifting, rather it owns steadiness and eternity. The one asking about true being must not rely on perception of outer reality nor lock his vision to it, nor orient himself on what is transitory. Rather he must view the Eternal and Everlasting which is beyond all perceived reality and which in all of it is the only true reality.[8]

Heraclitus, a contemporary of Parmenides, is famous for his dark and hidden sayings which become intelligible only when seen on the background of the experience of mystical union. Listen to some of those sayings:

True nature of being likes to hide itself. Reality is paradoxical as its true inner reality hides as well as reveals itself. The world is a fragmented world, but this is not the ultimate truth. Rather the fragments are deeply related with each other. Through their divergent actions the parts are joined into an underlying oneness. In the dependence of the opposites on each other and in their mutual interrelationship a deeper unity reveals itself. In reality the invisible but inherent harmony is more significant than the outwardly visible fragmentation. What tends apart reunites and out of the dissimilarity grows the most beautiful harmony. All is one. All becomes one and one in turn becomes all. All transformations reveal the one; while changing — all remains. All is alive, expanding and folding back into one. This is the deeper reality in the fragmented world.

Based on these statements by Parmenides and Heraclitus I say that at the origin of philosophy lies the conscious recognition of the distinction between everyday reality and true being. 'An ancient song we knew so long ago and held more dear than any melody we taught ourselves to cherish since' — this is true being, as we have heard it formulated in 'The Forgotten Song,' while 'the world we learned or taught ourselves *since* then' is the way we *ordinarily* perceive the world.

4.   The psychological evolution from unconscious to conscious unity

My claim that 'the experience of unity' also lies 'at the origin of psychology' is based on the observation that our consciousness of self evolves from unconscious unity at birth to conscious unity in the fully evolved state of mature wisdom. In the beginning, when we are born and within the first few months of life, consciousness still rests in all-encompassing undifferentiated unity. But life cannot continue indefinitely in this state of *all-oneness*. Consciousness begins to differentiate between self and non-self. In this transition the development leads out of *all-oneness* into *alone-ness*. Our exit out of our unconscious, undifferentiated unity in babyhood into the differentiating perception of a world full of polarities and conflicts corresponds to the expulsion out of Eden, out of paradise, in consequence of our tasting the fruit from the tree of knowledge, that is, as a consequence of our differentiating mind. 'Paradise' symbolically represents the un-differentiated unity consciousness of babyhood.

Human consciousness then develops along a series of stages characterized by successively wider horizons and increasingly sophisticated differentiation until, in late maturity, we *consciously* search and strive for an all-encompassing integration and *unity*. This has been practised and testified by the great philosophic and spiritual teachers of all ages and cultures. Quite often this *conscious search for unity* is triggered by an existential crisis which arises from the confrontation with the fragmentation and brokenness of the world. The sacrifice to be brought in this development is our own part in this brokenness, i.e., ourselves. When we are ready to *lose* ourselves we *find* ourselves — in all-oneness. The horizon then is radically unlimited: it widens to encompass all of heaven and earth. The path of development towards this end corresponds to ongoing progress from fragmentation towards unity. This end may appear to be unattainable, but it bears significance for us as individuals as well as for humankind as a whole in our search for meaning and orientation, as well as in any non-trivial, genuine integration of philosophy, psychology, and religion.[9]

We remember that 'The Forgotten Song' reminds us of our almost forgotten state of unity at birth when it says:

> Listen — perhaps you catch a hint of an ancient state not quite
> forgotten;
> dim, perhaps, and yet not altogether unfamiliar,
> like a song whose name is long forgotten,
> and the circumstances in which you heard completely
> unremembered.
> Not the whole song has stayed with you, but just a little wisp
> of melody,
> attached not to a person or a place or anything particular.
> But you remember, from just this little part, how lovely was
> the song,
> how wonderful the setting where you heard it ...

I understand this text as an expression of the experience of unity at the origin of our consciousness, an experience we all have shared but which we have forgotten. And what is the experience of mystical union if not a blessed moment of return into this eternal, timeless state of unbroken wholeness, interconnectedness, and unity?

5. The idea of unity in physics

We note, with some puzzlement, that not only the horizon of our consciousness, but also the horizon of physics and cosmology wants to converge on to a state of unity. What do scientists mean by that, and do these parallel developments in physics and in psychology point to deeper, hidden, meaningful interconnections between physical and mental reality? Let us postpone these questions until we have looked at what is meant by appeals to the idea of unity in physics and cosmology.

Let us begin with a few words about the holistic nature of physical reality:

5.1 Physical reality is an unbroken whole
What do we know about the true essence of reality? To our normal wakeful consciousness reality shows itself:
— as a continuously changing dynamic *now,* represented by the rich coexistence of a multitude of seemingly isolated objects, and;
— indirectly (via our memory) as a *now* which seems to have been preceded by what we call the *past.*

The signals of our physical measuring instruments are also interpreted by us in that same state of consciousness. Thus physics describes the 'now,' and cosmology describes the 'past' — if possible all the way back to an absolute origin in which all description comes to an end.

But now we realize of course that our *normal* state of consciousness is but a small cut out of much vaster possibilities of human consciousness. And *physical* reality is but a small cut out of much more encompassing aspects of the whole of reality. The whole of reality has a timeless and non-local quality which reveals itself in mystical union but not in our normal state of consciousness. Quantum theory describes the world as an unbroken whole which does *not* consist of parts. *Our consciousness* can analyse and cut the world, and it can do this in many different ways, for instance, into separate objects, into self and non-self, into spirit and matter, force and substance, space and time. *Our consciousness* can cut the cosmic event of quasi simultaneous genesis, evolution, and existence of the universe into a sequence of apparently separate moments, spread out along a coordinate called 'time': the disentanglement of an essential, true, and unbroken whole, or oneness, or unity. To our mind reality appears to consist of individual objects. But a 'scientifically complete physical theory [of isolated objects] is the (temporary) product of (philosophically) completely self-satisfied physicists.'[10]

5.2 The unification of space, time, energies and forces in physics
The underlying unity of physical reality reveals itself not only in this holistic quality of quantum physics, but also in what is called the unification of forces, of matter and energy, space and time.

The process of unification in physics reduces complexity. Each physical theory describes a wealth of phenomena on the basis of some unifying general principle, such as a law of force, or a principle of symmetry. Think, for instance, of Newton's principle of gravity: it permits a unified view of celestial mechanics and terrestrial mechanics; it unifies Kepler's laws of planetary motions and Galileo's laws of falling bodies into one common concept: the law of gravitation. Or think of Maxwell's theory of electromagnetism: it unites the totality of all electrical and magnetic phenomena into one unified concept. Or think of Einstein's special theory of relativity: it unites the concepts of space and time into a unified space-time, and it unites matter and energy. And take Einstein's general theory of relativity: it unites the space-time and the matter-energy density of special relativity into an inseparable space-time-energy continuum. In all of these steps of unification, two phenomena or two frames of perception that first appear to be of fundamentally different nature become one: when the two are regarded within the horizon of the underlying but initially hidden unity, then the unity reveals itself as soon as the appropriate mathematical and symbolic tools and viewpoints are discovered.

Along this path the entire complexity and wealth of forms in the material world has been reduced to the principles of gravity, of electromagnetism, and of the so-called weak and strong forces which govern the structures and reactions of atomic nuclei. But why should there be four fundamentally different types of forces? Why not one single root out of which the diversity of forces develops? This question lies at the core of most monumental experimental and theoretical efforts in physics and astronomy during recent decades.

It has been found that unifying views can be discovered when matter and forces are investigated at ever higher densities and temperatures. Types of matter and of forces, which under normal conditions have quite distinctly different characteristics, approach each other at increasing temperatures and densities. The highest technically realizable densities and temperatures are generated in collisions between elementary particles, for example in the high-energy accelerators at CERN near Geneva and at Fermilab near Chicago. In these experiments it is found that the properties and strengths of the forces of electromagnetism and of the nuclear forces clearly tend towards unification with increasing densities and temperatures.

On the basis of the tendencies towards unification observed up to this time we may expect that under most extreme conditions, which cannot be realized in any laboratories, we may envision an all-encompassing unity in such an absolute sense that all distinctions — between force fields and energy fields with their material manifestations, and between these and space-time — vanish.

The present situation in unification physics and cosmology can be envisioned as follows: We find ourselves in the crown of a tree, within foliage, blossoms, singing birds and humming insects. Looking from our position in the crown further into the tree we clearly see the leaves and blossom bearing twigs uniting into thin branches which then merge into thicker ones. Looking even further into the tree we see that all branches of the entire crown are supported by four thick principal branches. Following these further inward we notice that first two of these merge into one which, still farther in, seems to merge with yet another one. The farther we try to peek into the tree the more fragmentary becomes our view. We believe we see short lengths of just two or three thick branches. Sharply focusing on the angles of those branches we notice that they all seem to merge into one common origin. The boldest ones among us firmly believe that ultimately all branches merge into one single trunk. That would be the most satisfactory overall picture. The trunk itself we cannot see. But wouldn't the crown of the tree make more sense if there was one branch-bearing trunk?

The four principal branches are the four force fields of electromagnetism, weak and strong nuclear forces, and gravity. Electromagnetism and the weak force have been observed to merge, at high energy densities, into the so-called electro-weak force which, in turn, seems to merge, at still higher energies, with the strong force into the so-called interaction of grand unification. Attempts to merge this latter one with gravity are jokingly called 'theory of everything': The unity aspect of physical reality motivates the search for unification of the so-called Fermion, Boson, and Higgs fields with each other as well as with space and time in high energy physics and at the ultimate so-called 'Planck' horizon of cosmology. The concept of an all-encompassing unity of all physical reality cannot be examined in environments that are achievable on earth, but it plays the role of an absolute horizon in the application of physics to cosmology.

6. The concept of unity at the ultimate cosmological horizon

In cosmology the concept of unity corresponds to a world view according to which the whole rich diversity perceived by us in the present state of the world has evolved out of an original state of undifferentiated unity. This powerful and attractive idea implies the breaking of an initial perfect symmetry, and the disenfolding of diversity out of a beginning under conditions of most extreme energy density and temperature. Many astronomical observations, when considered on the basis of known physics and astrophysics, strongly suggest this scenario of evolution of the present universe out of a dense and hot original state which seems to have occurred approximately 15 billion years ago. In this scenario the universe today appears to us as a huge transparent sphere of finite size with a radius of 15 billion light years, containing billions of stars in billions of galaxies. At the distance of 15 billion light years we seem to be completely and uniformly surrounded in all directions by the surface of a hot and ionized gaseous spherical shell, a brilliantly shining opaque mist which acts as a boundary to the transparent part of the universe, thus representing our horizon of visibility. We cannot look farther because we cannot look into this opaque mist. It represents the edge of the transparent part of the universe. It surrounds us completely and uniformly in all directions. It is our horizon of the visible universe and, as an archaic mist, it shows us a young stage in the evolution of the cosmos, when it had an age of some hundred thousand years.

The scientific rationale behind this worldview stems from the following observation: when we look at cosmic objects at distances of, let's say, billions of light years, their light has been travelling to us for billions of years. The farther we look, the farther back into the past we see. Thus it comes that at distances of approximately 15 billion light years we reach back into a time when neither stars nor galaxies had been formed yet. Behind the farthest stars and galaxies and quasars we see the universe in a very young state in which neither stars nor galaxies nor dense bodies of any kind had yet accreted. Now, in that young state, when the universe was merely a few hundred thousand years old and had cooled from almost infinite temperature down to a few thousand degrees, *the heavens were aglow in almost perfectly uniform and pure golden light,* pervaded by a gas of ionized hydrogen that formed a hot, brilliantly shining, opaque mist.

Let us imagine the sight that this scene would have offered, had we been there to observe. We would have been immerged into that brilliantly shining but totally opaque golden mist. Then a dramatic spectacle would have presented itself to us. As the temperature dropped, along with the general expansion of the universe, the opaque ionized hydrogen gas of protons and electrons became a transparent gas of neutral hydrogen atoms. The mist became transparent everywhere. After one second we would have been at the centre of a transparent sphere of radius 300,000 km, after one year at the centre of a transparent sphere of radius one light year. Because, when the mist disappears everywhere simultaneously, then the light first must reach us to *show* us that transparency has occurred at other places too.

Therefore, the spectacle consists of this brilliantly shining golden mist seemingly racing away from us with the speed of light, leaving us in a rapidly growing transparent sphere, bounded by a seemingly still opaque spherical shell that recedes with the velocity of light. After one million years that brightly shining but opaque horizon would have been seen at a distance of one million light years. Today, approximately 15 billion years later, it surrounds us at a distance of 15 billion light years.

At this point I would like to read once more the central passage out of 'The Forgotten Song':

> Beyond the body, beyond the sun and stars,
> past everything you see and yet somehow familiar,
> is an arc of golden light that stretches as you look
> into a great and shining circle.
> And all the circle fills with light before your eyes.
> The edges of the circle disappear,
> and what is in it is no longer contained at all.
> The light expands and covers everything,
> extending to infinity forever shining
> and with no break or limit anywhere.
> Within it everything is joined in perfect continuity.
> Nor is it possible to imagine that anything could be outside,
> for there is nowhere that this light is not.

When I showed this text to a young Californian astrophysicist recently he was so moved and impressed that he went straight out into the nearest book store to pick up *A Course in Miracles,* the book which contains 'The Forgotten Song.' And that is exactly what I had done three years ago, after I had heard that text read to me for the first time: I immediately knew that I must get hold of that book.

But to those of you who are not astrophysicists I owe further explanations. Why can't we see this great and shining light which covers everything? The reason is that we are living in a huge and rapidly expanding universe. Because of the continuous general expansion of the universe the light from extreme distances is Doppler-shifted out of the visible range. If this were *not* so, if the universe were *not* expanding, and this archaic light were *not* shifted out of the visible range and out of the range of heat radiation, we would be totally surrounded by a sky background as bright and hot as the sun. It would keep everything everywhere at a temperature of several thousand degrees, preventing any possibility of condensation of water and of evolution of life. This was the situation for the first several hundred million years, or so, in the life of the universe.

As the range of visibility expanded farther and farther, with the speed of light, the brilliant horizon receded to distances where, in the expanding and huge universe, the expansion velocity of the ionized gas itself approached the velocity of light. Therefore, the brilliant golden light became shifted from the visible into the infrared and beyond. Today, at the age of the universe of approximately 15 billion years, the horizon has raced to a distance where the expansion velocity of the hot opaque mist at the horizon is only by about one millionth smaller than the velocity of light itself. Einstein's special relativity becomes a dominant feature in this situation. A relativistic dilation of wave lengths and oscillation periods by a factor of one thousand makes the golden glow, which is characterized by a temperature of three thousand degree Kelvin, appear to us as three-degree-Kelvin cosmic background radiation. The light and heat radiation fades away into the frequency range of microwaves. For this reason the night sky is black, and for this reason the radiation temperature of interstellar space is low enough to allow the formation of all sorts of molecules.

Hidden behind this hot curtain we may imagine an absolute horizon where the so-called curvature of space and time becomes so extreme that everything is merged into the one original, primary, primitive 'unity.'

The original state of unbroken unity, although hidden from our direct view, is to be thought of as an absolute horizon surrounding us in all directions in the farthest limit of the cosmos and in the beginning of time. At this invisible unity horizon all distinctions between space and time and matter and force come to an absolute end. There is no possibility of penetration through this horizon, neither in space nor in time: Our universe has neither a spatial nor a temporal 'beyond.' Whoever insists in mentally penetrating the unity horizon and 'looking' beyond always stays within our own universe. The unity horizon, at which all description comes to an end, cannot be penetrated. There is no physical sense in trying to think of a beyond. If we insist, in spite of that, to cling to time-bound thinking we can only say 'in the beginning there was unity,' and there is no time beyond that.

The concepts of 'time' and of 'beginning,' however, describe frames of our perception, conditioned by the limited abilities of our normal state of consciousness. That state, as we know, is only a small cut out of vastly more extensive possibilities of consciousness. Physics research approaching the state of ultimate unification touches on consciousness research, because consciousness is part of the whole of reality which any theory of everything would have to cover.

7.   The unification of physical reality and consciousness

The experience of unity lies at the origin of consciousness research, of religion, of philosophy, of psychology: today we note with puzzlement that the horizon of physics and cosmology, too, wants to converge on to unity.

Do these parallel developments point to deeper, hidden, meaningful interconnections? Does all this touch on the problem of unification between the realms of the rational and conscious with the realms of the spiritual and hidden?

In my own personal experience the unity of physical reality and consciousness reveals itself not in equations nor in a theory but in a way of thinking and living. There is no need to unify mind and matter, because they are a unity already. The sooner I realize this, and the sooner I adjust my way of thinking and living to this insight, wherever I believe I have encountered a conflict, the sooner I experience the healing power of this way of thinking. I used to believe that I am responsible for my way of living, for what I do, for my actions, but not for what I think. But the truth is that we are responsible for our thoughts, because our actions flow out of what we think.

One final remark: do not conclude that I have ever had an experience of mystical union *myself*. I have a friend who has had one *once*. To him it suffices for his whole life to remember and meditate that one experience. To me it suffices to know about it.

There is something that accompanies the experience of unity that is at least equally important: that is the experience of purification of the heart; that is the experience of total and unconditional forgiveness.

There is no genuine unity without purification of the heart and unconditional forgiveness. This leads to 'a vivid and overwhelming certainty that the universe, precisely as it is at this moment, as a whole and in every one of its parts, is so completely *right* as to need no explanation or justification beyond what it simply is.'[11]

Blessed are the pure in heart, for they *shall* see God.[12]

When your eye is single, your whole body indeed is light,
but if it is divided, your body is darkness.
Now if your whole body is light, having no part dark,
the whole is full of light
as when the bright shining of a candle fully enlightens all of you.[13]

Some people meditate every day but make no real inner progress. The experience of unity is just the beginning of the life of the mystic. Thank you for forgiving me for having said that.

References

1. *A Course in Miracles,* Foundation for Inner Peace (Glen Ellen), 1975, pp.445–47.
2. Abulafia, in Sholem, G. *Major Trends in Jewish Mysticism*, NY 1962: quoted from Wolpert 1996 (See note 5).
3. Plotinus, quoted from Wolpert 1996 (See note 5).
4. Buber, M. 1909: *Ekstatische Konfessionen*, 5th Ed., Schneider, Heidelberg 1984.
5. Wolpert, A., 1996: *A Meditation on Mystical Union Using System Dynamics*, Arlen Wolpert, 411 Franklin St. #1008, Cambridge, MA 02139, USA. Website: awolpert@world.std.com; http://world.std.com/~awolpert
6. Kolakowski, L. *Religion*, Fontana, Glasgow 1982: quoted from Wolpert 1996 (See note 5).
7. Heschel, A. *God in Search of Man*, Farrar, Strauss & Cudahy, NY 1955: quoted from Wolpert 1996 (See note 5).
8. Weischedel, W. *Die philosophische Hintertreppe: 34 grosse Philosophen in Alltag und Denken*, dtv, Munich 1975.
9. These remarks are based on: Staindl-Rast, D. 1985, *Science and Religion*, lecture, Cortona Week of ETH, Zurich, on *Science and the Whole of Life*; cf. also: Fowler, J. W. *Stages of Faith*, Harper & Row, NY 1981.
10. Kyprianidis, A. & Vigier, J.P. 'Quantum Action-at-a-Distance: The Mystery of Einstein-Podolsky-Rosen Correlations,' in *Quantum Mechanics Versus Local Realism, The Einstein-Podolsky-Rosen Paradox*, Plenum, NY 1988.
11. Bucke, R.M. *Cosmic Consciousness*, Innes, Philadelphia 1901.
12. Matthew 5:8.
13. Luke 11:34-36.

Biology and Gaia

A Holistic Philosophy of Nature

GLEN SCHAEFER

Professor Glen W. Schaefer was Director of Ecological Physics Research Centre at Cranfield Institute of Technology. He graduated in Mathematical Physics at the University of Toronto and took a Ph.D. in Quantum Field Theory under Sir Rudolf Peierls at the University of Birmingham. This led to ten years of research into the mathematics of nuclear reactors during the early days of development of atomic power in Britain. His second career in biology and ecology applied the physical sciences and technology to ecological problems encountered by governments and crop protection agencies worldwide in relation to 'pest control' on a wide variety of crops. His research made possible, for the first time in the West, the successful replacement of chemicals in large scale crop protection by the use of natural biological organisms. He latterly attempted to formulate and apply an alternative metaphysics which takes the spiritual and mental universes as primary and the biological and physical as derivative. He died in 1987. This lecture was delivered at the 1981 Mystics and Scientists conference.

There have been about forty thousand scientists working in a detailed way studying nature for a hundred years. That is four million man-years of detailed investigation! An impressive picture has emerged, or so it seems. It is an evolutionary picture. It starts with a Big Bang; we can go right back to origins, it says. We go through a cosmogeny down to life, to DNA molecules, through Darwinian processes all the way up to Man. It is largely a deterministic and mechanistic picture. There are, however, a handful of scientists who see major flaws in this picture. I believe it demands a completely new science, an expansion of existing science almost out of all recognition. I would like to outline a few of those catastrophic phenomena that are now forcing themselves on the thoughts of the most thoughtful. I shall try to highlight some of the disturbing facets in cosmogeny, then in biological evolution, then in Man — dealing with his food — in insects, and in the mind-body relations and higher dimensions slightly at the end. How can I attempt an interpretation of so infinite a theme? Of course, I cannot — not alone. Together, we just might get somewhere if we are specially favoured. That is, a seed might just grow.

You might like to know a bit of my background in order to assess whether what I'm saying has any foundation or not. You'll probably think it hasn't, even if I do tell you my background. I'm Canadian, still am. I've been in Britain over twenty-five years; it's my chosen place to work. Since the age of five, I've been greatly influenced by the metaphysics that my grandmother was able to teach me, then through my parents in the early days. I had quite a few spiritual healings over the years. At age 10, I became intensely interested in atomic physics and mathematics. At age 12, in biology, particularly birds and insects and migration. By the age of 15, I knew I would have several careers. I have now gone through ten years as head of research teams in atomic physics, quantum mechanics, nuclear reactor theory, then into biology for the last twelve years, mainly concerned with insects, crops, pests and meteorology, ornithology, spraying and so on. Now, at the moment, I have ten Ph.D. students and I work mainly in these ecological subjects. I'm a consultant to quite a few organizations dealing with the protection of large areas of the earth. I'm about to go to China on a lecture tour and consultancy to deal with major problems of their crops. Faced with such responsibilities, which I cannot possibly carry out, and being a mere human at this early stage in history, I'm looking for something to help me to answer the questions accompanied by millions of dollars. These are huge problem areas. Our crops are under severe attack, our food sources are very fragile now and I'll tell you more about that later. That's when we come to Man and the environment. First, however, I want to start with the universe.

Against all the odds

I want to give you a few metaphors and a few linkages to help form some notion of what might be required of future science. I begin by noting that there's been a long-term paradox, which most scientists do not face because they are too interested in their work. It is that our western paradigm in science, and therefore in all of your lives, is atoms and chance and the laws of physics, mechanistically evolving from the lesser to the greater, up to Man — with mind added as extra. To me, and this is where the paradox is, we also know, as thinking humans, that the picture that's come out of four million man years of work is only a picture. It is a result of great mental activity; concepts, intuitions, ideas. It is a result of mind, but the picture seems to be mechanistic. The universe seems to affect us through the senses and hence the mind, yet we know that mind must be working the opposite way around, possibly projecting as: 'I believe the universe and the senses.' However, I don't wish to push that point. Instead, I want to talk about catastrophes now appearing in the paradigm.

To date, western science has chosen to go from the 'outer' to the 'inner' as the prime direction. This paradigm has led to neolism — we are 'specks on a speck on a speck on a speck in an enormous universe,' and therefore there is no meaning, only chance. We must fight to conquer nature by matter. We use chemicals in medicine to attack our bodies, we use chemicals in our environment to attack insects and so on. This is what we do as humans, and we know that, yet we don't face up to it. Following this route of natural science to great depths, in four million man years, is producing for the very first time extremely clear views of a catastrophe for the current scientific paradigm. I believe this catastrophe will lead to liberation for Man.

The Big Bang theory looks at evolution from the cosmos down to the level of the Earth. This surely is the ultimate of the paradigm. We have now found a materialistic origin for all of us and our minds. We start with an explosion of light at time zero, leading within a few microseconds to corpuscles, elementary particles, within a rapid expansion of the fireball. This cools and condenses out more atoms. The atoms collect into stars, stars collect into galaxies, planets emerge out of the gas around stars and life gets going on certain planets. The picture seems quite perfect. Moreover, you probably believe in it, because we are so materialistically orientated. Yet, in order for this picture to work, there has to have been an amazing sequence of coincidences. Let us now consider them, and in doing so, uncover the first evidence of catastrophe for the paradigm.

First, there is exactly the right expansion rate for this expanding universe. If it went any slower, the matter would have collapsed long ago — the universe would have sunk back to a black hole. If it went any faster, galaxies could not have condensed out and you would have not been here. Life would not be possible. If you look in extreme detail at the theory you will see that, in changing the expansion rate in the early moments of the universe by one part in a million-million, it goes either too fast or too slow. It is *precisely* the right expansion rate. Now that is a coincidence. It is thought-provoking.

Next, there is too much order in this 'universe of the fireball'. The fireball should have been — mathematically-speaking — 'chaotic' in that most extreme of high temperature. Out of chaos came the universe. Yet, when we look around with radio astronomy instruments we find extremely smooth isotropic conditions in all directions. There is almost no irregularity in the overall density of the universe. The theory has great difficulties in trying to fathom out how you can have such regular order out of chaos, out of supreme randomness, out of billions of degrees centigrade at the start. Well, maybe we will find out some day.

Consider the nuclear forces, the forces that hold protons and neutrons together at the centre of every atom of the universe. If you change those nuclear forces by two percent, you will not be able to have life — there would not be any heavy elements and life would not be possible. Two percent! Well, we explain it away with another 'coincidence'. Whoever was doing this thing, got it just about right. Electrostatic forces in the electron-to-proton mass is exactly right within one percent. In order to have molecules, we have just the right size to form complex molecules and therefore life. Got it just right again!

Each one of these is a coincidence, but you all know from your own human experience that multiple coincidences are amazing things and beyond the usual definition of the word. These sets of coincidences are very disturbing to ten or twenty people in astronomy, who actually think of what they are doing outside the extreme intelligence of their theories. It is quite clear that such coincidences mean that the end of the evolution is in the beginning. The paradigm cannot work. We cannot be going from the lesser to the greater.

To salvage the situation, these few people have forwarded a new principle, outside of physics, in order to hold this evolutionary picture together at all. It is called the 'anthropic' principle — the 'Man-centred' principle. No one understands this principle, but it's absolutely essential to have it. We need it in order to reason out that the end should be in the beginning, to overcome these amazing coincidences. It is extremely perplexing. Most scientists will not face such a conundrum in the paradigm.

Let us come down to life now, away from the universe for a moment, the apparently inanimate universe. Let us look at life in space first.

We all know that there is dust between all the galaxies throughout space. All telescopes show that. It was thought for a while that those were just atoms. Then it was discovered by Fred Hoyle and others that they were organic molecules. Now, there is evidence that in fact they are not simply organic molecules but living things filling the universe. The dust is living!

Before even getting to the living stage, one must question the probabilities that those molecules could have been formed in space by random events: hydrogens, carbons, and so on flying around, meeting, sticking, according to the laws of quantum mechanics Those numbers have been calculated and the probability of getting even a single one within the entire 10,000 million years since the Big Bang, within all 10^{20} galaxies, show it is not possible. The probability of even getting a *single* such event is about 10^{-200}. These probabilities are essentially zero. They are so 'zero' that you have to consider them as zero. Chance alone is not capable of producing these molecules, and it's driving some major scientists into a new state of consciousness, Fred Hoyle being one.

The fact that these things are not just organic molecules, but living organisms means that chance events are even more unlikely. One must conclude that there is no way that the Big Bang theory can hold. Chance alone is not possible. There must be an impressive information source or intelligence behind this and there is no way that the improbabilities could ever have been overcome in such a short time as 10,000 million years.

When Hoyle was interviewed on BBC radio just three weeks ago by John Maddox, he was asked: 'Do you mean to say, in the words of Adam Smith, that you believe in an invisible hand?' Hoyle answered, 'Yes, I do. I'm now going to devote the rest of my life to discovering the nature that definitely exists.' By 'nature,' Hoyle meant this intelligent source, this intelligent background, this principle whatever it is. The near-zero probability of random events creating life in the universe is driving some biochemists to extreme positions like creationism. Hoyle's idea that there is an intelligent background before evolution can ever get going, impresses me tremendously.

Breathing life into the clay

If you look very closely at the Darwinian approach to evolution of life on earth, you are not too surprised that there is a tremendous amount of evidence for natural selection. We all know that you can have giants of this and pygmies of that within a species. However, there is no evidence, it seems, for the *evolution* of species. None. That is a very strange thing.

There shouldn't be any evidence, by the way, if the Big Bang theory and chance hold. We've just discussed the impossibility that chance could have evolved things by mutations or whatever in the time available. Something has got to give. Either you have an eternal universe, or you have one that is 10,000 million years old with a creative, intelligent principle. I believe that. All the sources of information that I'm using are from people who have worked at least twenty-five years and are at the head of their fields, and are in print. If you look into those positions, you will find that there is no evidence for evolution and they say so. We believe it because of our education. It is pure illusion like almost everything else in education. We don't know what else to do. What do we do?

Genes. DNA. Surely we've got something mechanistic here? I think not. It is only when you get into the details of the materialistic picture that you see it can't possibly be so. You can't deny the evidence that something is happening, but the story that goes with it through chance and mutation has had to be seriously challenged as well.

There is a wonderful little book, a tiny book, costing about fifty pence or so — amazing value, the best book in the universe I should think per pence! It is titled *Homology, the unsolved problem,* by one of the greatest Darwinian biologists, Gavin De Beer. It is well worth looking at. It is full of shocks and the greatest shock come when you consider DNA and genes. The story is even more catastrophically collapsing if you care to look at it. For example, *Drosophila*, the little fruit fly, which has had more genetic experiments done on it than probably anything else. It possesses a gene called 'eyeless' (genes have names, you see. They are associated with eyes or ears, or whatever).

The eyeless gene is recessive, if the fly inherits it from both parents, it deprives the animal of eyes. However, if you continuously inbreed eyeless *Drosophila* after several generations you get offspring with eyes. If you try to find out whether the eyeless gene, which was identified at the start, was still there, it is, and it is unchanged. So, it wasn't controlling an eyeless eyes condition at all! What happens is that other genes apparently deputize in the absence of the normal genes for eyes. They deputize if you haven't got a normal gene to prevent the eyeless condition. Why? How?

I conclude, and this is my own personal view at this point, that genes and all the other biological mechanisms in our bodies and every other body are really servants. They are servants that create a reductionist pathway to serve a higher pattern, a pattern that lies possibly behind 10^{-200}. All over the universe there is a higher pattern pushing its way through. The mechanisms, which mechanistic biology thinks are the masters, are really the servants. They are under instruction, so to speak, to create pathways. If we blur a pathway or rub it out between here and there, another pathway is found. The genes are extremely pliable, the mechanisms are extremely pliable and in a moment you'll see how pliable insects are.

Around 1900, before Darwinian things got going, biologists, geologists and others were looking at the universe. They tried to explain things in terms of *archetypes,* a kind of platonic idea. The biologists didn't know about genes and things of that sort. Darwinism put paid to all that kind of idealism and idealistic thinking.

To speak metaphorically about the eye for a moment, light seems to me to be a supreme archetype that must be witnessed. An eye must be manifested in some form, not necessarily materially. A pathway must be produced, a principle of least action, something governs from beginning to end.

Looking at all these coincidences — the 10^{-200} and pliable genes — the universe and life, I believe, are not evolved from the lesser to the greater, but rather the mechanisms (the lesser) are pliable and plastic and adaptable. They respond to a higher life principle of some sort, to manifest systems of interlocking phenomena. Evolution is really unfoldment governed by a higher dimensional ground than matter or the human mind.

Our paradigm of the deterministic Big Bang concept, which we all suffer from and work upon most of the time although we may not believe, is still in force in most of our lives. The paradigm says, if you want to protect your crops when insects seem to be around in great number, get out and kill them. Man has the right to control, control is the concept, mass kill is the method, chemicals are the agent. But, what does nature say to this approach?

I'm intimately involved now in this particular area in many countries on five continents. Wherever crops are grown in large amounts, an insecticide is used. This happens nearly everywhere, particularly in the third world and in China. Surprisingly, more insects appear following insecticide use than before it, and the more insecticide you use the more insects you have. This is another paradox.

Our stupidity is in believing that we have the right to control and to kill in the process. Our first reaction to an insect is to step upon it unless it happens to be a rather beautiful butterfly. We do not realize that everything must fit together. If you have 10^{-200}'s floating around the place, actions must fit together perfectly, there's no chance about it. It is all nearly perfect. So, do you wonder why with more insecticide you get more insects? It is a billion dollar problem. We are down to the last few insecticides due to natural enemies becoming wiped out faster than the so-called pests. Resistance is setting in, which you've heard about. The answer is that it is exactly like treating your body with chemicals. The side effects are the same in the body of nature, the same type. This is now gradually being realized largely through my work. Resistance cannot be predicted, none of these things can be predicted. We have no control over our understandings of this, but nature is fighting back. It has to hold together, for these 10^{-200} reasons it has to hold together,. It has to go from here to here, things have to grow because the whole thing is intertwined in an ecological network.

If you start rubbing out this pathway by blasting insecticide and seeing all the bodies falling down, you think you've done it. However, Nature always reaches the same goal again, except in the meantime more insects have come along to show you as a mirror to your own theories that you don't know what you're doing. It is disastrous.

There are indications now that resistance to the latest chemical insecticides is developing in many insect species. It takes twelve years on average to find a new chemical, plus millions of pounds, and the insects can evolve in a matter of three or four years. That is through natural selection — unnatural selection in this case, by excessive mortality. We have realized, particularly through ecology, that we are faced everywhere with huge forces nearly in balance. By applying small forces, we can tip those huge forces and we have huge effects. This is because we don't realize how nearly perfectly balanced they are. A slight shift of billions on both sides gives you millions.

There is a mirror principle here. I first stumbled into this when I was trying to think how to get away from insecticides, or any of the other things: sterile males don't work, sex pheromones don't work and so on. Nature finds a way around it. It is so pliable. We can't predict any of these things because genetics is so variable, because meteorology is so variable, because insect migration is so extensive. What are we going to do about it?

First of all, let us learn what does go on. Almost all science takes place indoors: physics, biology, and so on, which is a crime. In my view, you can't learn anything indoors about the universe, because you make your little tiny universe and you follow it for the rest of your life. Nothing is learned or we wouldn't be in this situation of realizing that our paradigm has got catastrophic effects all through it. We can stop killing to a certain extent. I've learned how to cut the chemicals by a factor of 10 for the same job, but that's not enough. It is homeopathic, on the way to homeopathy, which is fine. That was a good solid early base to medicine, probably the most solid thing that ever happened in medicine. All these wonderful things we know about in medicine are really hygiene.

A visual picture

I want you to imagine a map of eastern North America, showing the arboreal, evergreen forest. You might be able to see the North American sea coast here: Newfoundland, Nova Scotia, New York and the Great Lakes. White patches would show you parts of the forest that are severely defoliated by moth larvae eating the buds. There was a natural cycle to this defoliation. An outbreak of moths starts up in a small area. For a while the population is more or less steady, then it explodes. Suddenly, there are two million acres of forest defoliated right across from central Canada to the East Coast. Then it disappears, because the trees died, were blown down, or burned in forest fires. Young trees come up.

Without this cropping by this one species of insect, we would have no forest at all. It is the way the old trees give way to the young, which are always there but can't make their way when the overstorey is too filled in.

If you happened to be living shortly after the war in the middle of these defoliating areas, you thought, 'My God what a catastrophe! We need the timber to produce Sunday newspapers and toilet rolls.' The forest industry said, 'We've got chemicals from the war now, let's use them.' So they sprayed and the plague went away. However, you've had to spray every year for the last twenty-five years in order to keep it that way.

Rachel Carson used this particular situation as an early example of something which is true everywhere on earth now — not just in forests, but in every major crop everywhere on earth. Britain is one of the least effective countries in this way. We have been holding back a great deal on insecticides, but the pressures are on — the materialist paradigm pressures, commercial pressures, competition, everything that goes with it. That is the situation everywhere. It is so bad even the physicists are involved in biology!

I found great help from artists in trying to find a way out of the reductionist mist in dealing with our environment. It is only early days, but there have been very powerful hints to me. William Blake was the first.

Two or three years ago, I went to the Tate Gallery to the great Blake exhibition. There on the wall among the three hundred exhibits was the ghost of a flea. The ghost of a flea! That is what I wanted. I wanted to know the nature, the mental and spiritual essence of an insect. You can get so far by using your eyes, but then you need new eyes and here was a drawing, a painting of the ghost or spirit of a flea. It was very revealing to me. It stunned me because the caption was: 'The ghost of a flea.' You expected six legs at least, but no. It is in man-like form, demonic, but man-like and that is the clue. That has altered my life.

Blake was reported as saying by a friend that he saw this image at the end of his living room, and he started to draw it. The image spoke to him while he was drawing; the flea spoke to him. The flea sucks blood and it has pincers or pricks for getting the blood, but the form is man-like. The flea said to Blake: 'Fleas are inhabited by the souls of those men who by nature are excessive with greed.' You see bloodsucking is greed. It is living off the life of others, or other things, or other parts of the universe. That image really has helped me very much indeed. The mirror of nature, as I have said, shows us back immediately our mistakes and multiplies the problem. The problem is not nature, the problem is us; that is what Blake is saying.

Then there is St Paul in Romans 8. It is an amazing paragraph to read after seeing Blake. He says: 'We know that the whole creation groaneth and travaileth in pain, (just like us) together until now, for the earnest expectation of the creature waiteth for the manifestation of the sons of God.' J.B. Phillip's translation I think is even more beautiful: 'The whole creation is on tiptoe to see the wonderful sight of the sons of God coming into their own.'

Redemption is the essence, not chemical control. Not will. Not concentration. Not human power. Humanity is not the source, but through our changing, the universe will change as well. The flea will be seen by us as it is, rather than expressing our greed.

How do we perfect our consciousness? I think that this is the necessity. The ghost of the flea. How do we see the flea as a practical problem facing all of us because of our food supplies? It's also a theoretical problem, but the practical one is going to get us in the end. How to perfect consciousness?

We have to demarcate, don't we, between the true thoughts and concepts on the one hand, from the evil and false concepts and thoughts on the other. How do you distinguish? Holism is required, the holistic philosophy of nature in my title. The necessity for it, I hope I've demonstrated to a certain extent. Holistic comes from the word *hole* which is the root of 'holy.' It certainly has something holy about it, as it comes from the same word root as *hale,* an old English word meaning 'healthy.' It is through health of ourselves and our natural environment I believe that we will learn about ourselves. We will be faced with problems of health, and we are ready in our bodies and the body of nature. The holistic philosophy has to answer the urgent questions of health, which are directly related to the holy, the holism, because if we don't get the whole we are nowhere.

As well as his principle, for which he is famous, Heisenberg said, 'Demons can be let loose and do a great deal of mischief.' To put it more scientifically, partial orders that have split away from the central order, or do not fit into them, may have taken over. Demons can come out of science as much as beauty and usefulness, demons like bombs and so on. I personally am not against atomic power in the slightest, but how do you demarcate between the good and evil uses of things? The new science, the new religion, must help us with this.

The new science will have to be *lived* not just experimented with. It needs always to be testable, applicable to our health, to the health of the environment. That is where it's going to be rooted, that will give it reality and realism. It can't just be a thing of the mind only. Even though the environment may well be a projection of the mind, it is showing us through the mirror where we stand. So, let us all see if we can alter our vision in some way by expectation, hoping that we shall see the amazing central intelligence.

The Environment Now
and the Gaian Perspective

JAMES LOVELOCK, FRS

Professor Jim Lovelock was born in 1919, educated at Manchester and London University, and now lives in Cornwall. In 1961 he took up a post as Professor of Chemistry at the University College of Medicine, Texas, where he worked with NASA, helping formulate the experimental research testing for life on the Lunar and Martian space probes. This led to his investigating the stable conditions that have made life possible on Earth for 3,500 million years, which could only have endured if there is a very sensitive self-regulatory mechanism similar to the way our physical bodies adjust to outside temperature fluctuations. This line of thought led to the 'Gaia Hypothesis' which considers the Earth as a single living organism. One of the most hotly debated topics in scientific circles, his theories are described in a number of books: Gaia *(1978),* Ages of Gaia *(1988) and* Gaia: The Practical Science of Planetary Medicine. *He is now an independent scientist working with industry and is also heavily involved with promoting research into the Gaia Hypothesis. He has been a Fellow of the Royal Society since 1974 and is also President of the Marine Biology Association. This lecture was delivered at the 1989 Mystics and Scientists conference.*

I think everybody here has heard of Fritz Schumacher. He was a very great man and although few would have read his works in full, most of us remember those very telling one-liners like 'small is beautiful' and 'act locally but think globally.' For me these phrases encapsulate a small but important part of his philosophy and during the past twenty-five years I've tried to do science in the way that he recommended. I take the advice to think locally but act globally most seriously and I've practised global science as a family business, if you like, done at home. It was from this kind of science that the idea of Gaia grew, not from the kind of science that seems to be a vast remote and potentially dangerous sort of activity.

I've lived for the past eleven years in a little cottage on the River Carey in west Devon, just on the border with Cornwall. My laboratory is a room built on to our cottage and the nearest house is about half a mile away. The nearest village is St Giles on the Heath, which is two miles. I'm afraid the title of my home is a bit pretentious, Coombe Mill Experimental Station, but it's very necessary. You see, if you're doing science you might want to order a pound of plutonium or a kilogram of potassium cyanide and if you tried ordering that from 11 Acacia Villas, Finchley, it won't work. However, if you send off from an address like 'Coombe Mill Experimental Station,' you can order anything!

So, my home and the laboratory is in the depths of the country a long way away from everywhere. This is necessary for me in my science, because I'm measuring trace quantities of things in the atmosphere and I don't want any pollution, but it also no doubt protects my neighbours from me. I've always thought that science is something to be done at home like writing, painting or composing music. After all, there isn't anything unusual about an artist doing his creative work in his home. Indeed, the very idea of an artist painting portraits in the department of fine art at the local university, or of a novelist commuting daily to the Institute of Writing, both seem utterly absurd. So, shouldn't science be done at home? Now I don't mean to be prescriptive in saying that and I recognize that much of science needs teamwork and a large scale of operation, just as there are teams of artists and craftsmen who build cathedrals or who form orchestras. It's just that science in modern times has been no place for individual thinking or experimenting. What I want to try to show is that this lack of individuals with the time and opportunity to wonder about and explore the world has grievously weakened our understanding of the natural environment.

Part of the problem arises because most of what people call, or think of as 'science' is in fact technology. A scientist in my opinion is somebody who has the time and inclination to *wonder* and who then expresses a personal view of the natural world, has theories and ideas that can later be tested by the accuracy of their predictions. There are very few of these and most scientists are really, if you think about it, like highly talented individuals who write the jingles and make the glossy pictures that advertise commercial products. Now, none of those people would claim that their work is art, still less that they were free to write or paint as the spirit moved them. Their position though is no different from that of most scientists in the world today.

My own special interest as a scientist arose from wondering about the earth and life upon it. My interest developed by just *wondering* — just standing and staring. It has become a new theory of evolution and it's a theory that sees the evolution of the species of organisms and the evolution of the material earth. The theory sees the air, the oceans and the rocks, not separately as in the division of science into biology and geology, but as a single tightly coupled process. The self-regulation of the climate and the composition of the air, the oceans and the rocks are then all seen as what the philosophers call an 'emergent property' arising automatically from that tight coupling of living organisms and their environment. This regulation proceeds without any foresight, planning or mysticism. It is almost as if the earth were a living organism and of course, as you well know, on the advice of the novelist William Golding I called the theory 'Gaia' after the Greek name for the earth, or the earth goddess.

A new view of the earth

To understand this new view of the earth, we have to take a very different approach from that usually taken by scientists when they investigate living organisms. We have to use what physiologists and engineers call a 'top-down view.' It is quite simple really, a top-down view is just what it implies: a look at the earth from space as a whole entity — the kind of view that astronauts first saw as they swam in space above the earth. It's the approach that physiologists use for living organisms and the approach that engineers use when they want to study a computer. An engineer wouldn't dream of vivisecting a working computer, he would try it out and ask it questions. In contrast, the bottom-up view is the usual ground-based view of the biologist. Take the system apart, vivisect it if necessary, consider the parts and then you will know entirely how it works. This reductionist view is unfortunately sometimes necessary in science, but both approaches are important and for the earth we have had far too much of the bottom-up, reductionist view and too little of the view from above. For example, I think most scientists and many others have read that splendid book of Richard Dawkins *The Blind Watchmaker*. It's about how life appeared and evolved and it uses the metaphor of Paley's watch — the watch that the divine Paley found lying on the beach and wondered how it had got there. Richard Dawkins is concerned entirely with how the watch appeared and how its parts assembled themselves. He spends no time wondering about the watch itself and what it's there for, or how it works. Watches are like living organisms, they exist and do their thing and that is all that need be known about them. That is why, of course, if you look it up the Dictionary of Biology you will find no definition of 'life.' Life everyone knows about, it's a given and biologists are not particularly concerned with it. Now, Richard Dawkins' questions about how the watch or living organism originated and evolved are perfectly legitimate in science themselves. However, they ignore the fact that the behaviour of complex systems like living organisms, Gaia, or even some mechanical contrivances can never be predicted from a mere knowledge of the properties of their component parts. In other words, the bottom-up approach just won't do. Even with a single self-regulating system like an automatic pilot on a ship or an aeroplane, it's much more useful to analyse it by seeing how it works than it is to take it apart and try to understand what all the bits of it do when they're working.

The top-down approach isn't new, I'm not offering anything novel at all. It's the usual approach taken by inventors, by engineers and by physiologists. It just happens to be out of fashion on mainstream science. This is very odd, because before the nineteenth century scientists were quite comfortable with a physiological view of the earth. One of them was a man called James Hutton and he's often been called the father of geology. In 1785, he gave a prestigious lecture before the Royal Society of Edinburgh. It's quite a long time ago, just over two hundred years, and in this lecture he said: 'I consider the earth to be a simple organism and that its proper study should be by physiology.' Now, James Hutton's wholesome view of the earth was quite acceptable to most scientists at that time, but it was discarded in the nineteenth century. It wasn't discarded because it was thought to be wrong, it was because the growth point of science changed from the holistic in the nineteenth century to the reductionist one. This was very understandable because in the nineteenth century people were going out exploring and exploiting the earth and they were gathering facts faster than they knew how to digest them and to make a whole picture. This was done independently by both the earth scientists and the life scientists. For biologists, in their gathering of bits of pieces of organisms, there was Darwin's great vision of Evolution of the Species and organisms by natural selection. On the other hand, for the geologists (who rarely spoke to the biologists) there was the wholly independent theory called 'uniformitarianism'. It actually originated with Hutton and said the evolution of the material environment was simply a matter of physical and chemical determinism. So, both the earth and life sciences were moving off in their own way, away from Hutton's vision of a whole earth.

It wasn't long before the earth and life sciences started to divorce and in the nineteenth century this divorce was inevitable. Not only was there a rapid increase in the supply of information about the earth as exploration and exploitation developed, but the techniques for looking at organisms were very different from those for looking at the ocean, the air and the rocks. It also must have been an amazingly exciting period in which to have been a young scientist, because there were new discoveries coming in all the time. However, nobody had any time to stand and stare and look back and take a broader view, or try to keep alive Hutton's super-organism. There's nothing surprising in that, it's very human and very natural. However, what I do find surprising and remarkable is that the subsequent division of science into earth and life sciences and later into even further separations, has persisted right up until the present day. The reason for the endurance of this division is, I think, a mutual acceptance by both geologists and biologists alike of the anaesthetic idea of adaptation. You see, a great part of the Darwinian theory that inspires biologists is the notion of 'adaptation'. Biologists have assumed that the physical and chemical world is evolving according to the rules laid down in the geology department at their university, which although interesting need not concern them in their quest to understand the evolution of organisms. They all think that one day they'll have time to go and read about it, because it is interesting, but in practice, as you well know, one never does such things. Biologists are comfortable with the notion that whatever change occurs in the environment, the organisms will simply adapt to it.

In a very similar way, the earth scientists are quite happy to accept this idea of adaptation without question, because it frees them of any need to constrain their models of the earth to account for the needs of the organisms. After all, the biologists tell them there are organisms that live in hot springs at a hundred degrees centigrade and that there are other organisms living on ice at freezing point. That is a wide enough range for any geophysicist who is a climatologist.

Adaptation is a very dubious notion in the real world, to which the organisms are adapted. The environment is determined also by a neighbour's activities rather than by the blind forces of chemistry and physics alone. In such a world, changing the environment becomes part of the game and it would be absurd to suppose that organisms would refrain from changing their environment if by so doing they were enabled to leave more progeny. I am sure that in his time Charles Darwin would not have disagreed with this. However, we must not forget that at that time he did not know, as we do now, that the air we breathe, the oceans and the rocks are all either the direct products of living organisms or else have been greatly modified by their presence. In no way do organisms just adapt to a dead world determined by physics and chemistry alone, they live in a world that is the breath, the bones and the blood of their ancestors and that they are now sustaining.

The Russian scientist, Vladimir Vernadsky, saw this separation of the earth and life sciences was much too extreme. He was the father of the modern science that is called biogeochemistry. I'm afraid it's a bit of a maledictive combination, but we scientists are liable to do that sort of thing! He and his successors, like Hutchinson and Redfield, recognized that life and the physical and chemical environment interact, that gases like oxygen and methane are natural biological products. Where they differ from geophysiologists or Gaian people like me, is that they still accept without question the dogma of mainstream biology, which is that organisms simply adapt to the changes of their material environment. What they failed to see were the consequences of adapting when it happens in an environment that is adapted by the organisms themselves. Yet, Vernadsky's worldview has been developed and expanded into what is now called co-evolutionary theory. It is present in biogeochemistry and biogeophysics and in the philosophy behind the very latest of the prestigious global programmes, 'The International Geosphere Biosphere Programme,' which the governments of the world have all joined in together to try and understand the earth. Good luck to them.

Emergent properties

The fundamental flaw of this mainstream approach is that it is still largely a bottom-up view and it cannot deal with emergent properties, properties that would lead to self-regulation of the chemical composition of the climate of the earth. Most importantly, it does not allow at all the possibility that the earth is alive in any kind of sense, not even as a physiological system. So, you may well ask, what is going on? Well, like co-evolution, Gaian theory rejects the apartheid of Victorian biology and geology, but it goes a bit further. Gaia is an evolutionary theory, one that sees the evolution of the rocks, the atmosphere and the oceans, the evolution of the species of organisms as a single tightly coupled process. A dynamically-evolving system of this kind shows emergent properties; that is, there will be more to the sum of its parts than you would expect just from considering them separately. For Gaia, the emergent properties would be the regulation of climate and chemical composition.

This kind of system is notoriously difficult, if not impossible, to explain by cause-and-effect logic. It is something that practising inventors like me know to their cost. It's very easy to invent something that is a cybernetic system but it will take two or three years to explain to your colleagues. It is doubtful also if the fashionable and trendy use of falsification tests, which are so valuable in theories of physics, are really applicable to systems like Gaia, or indeed living systems generally. If you doubt that, consider for example the problem that would confront a friendly alien from outer space who is unfamiliar with life on earth and who wished to find out if a Lombardy poplar tree was alive. Now, these trees are all males and hence can only be propagated by cuttings. Ninety percent or more of the fully grown poplar tree is dead wood and around it is dead bark. All that's alive is just a thin skin of living tissue around the circumference of the wood. In many ways this is a wonderful analogy of the earth itself — there's just a thin skin of living tissue around the circumference of the earth. Above it is a dead atmosphere that protects that delicate skin and below it are dead rocks just like the tree. What is so interesting is that the dead matter above the tissue of the tree, the bark, and the wood beneath — the air above us and the ground below — both were almost entirely made from material that was once part of living organisms.

From the circumscribed and dogmatic viewpoint of neo-Darwinism 'life' is simply defined as something able to reproduce and the areas of reproduction are corrected by natural selection amongst the progeny. This definition seems reasonable but if it's dogmatically applied then neither the poplar tree I just mentioned, nor a grandmother, nor Gaia are alive; they're dead. The neo-Darwinist, John Postgate, made a scathing attack on Gaia in an article in the *New Scientist* last year entitled: 'Is Gaia too big for her boots?' It was an amusing article and well worth reading. In it, he implied that Gaia was merely a fanciful restatement of the balance of nature, something known to all biologists right the way since the nineteenth century, indeed he said his schoolmistress had taught it to him. He quoted Huxley who had said that green plants use carbon dioxide for growth and excrete oxygen and animals use plants as food and excrete carbon dioxide and therefore there is an automatic balance that sets the fluxes and the abundance of oxygen and carbon dioxide in the air. That is all you need to know, you don't need to know anything about Gaia.

John Postgate's article read as a devastating put-down for Gaia and I ought to have been dismayed. I was not, because other equally strong critics of Gaia, the eminent geochemists Jim Walker and Dick Holland, both firmly denied John Postgate's views. They claimed that the levels of both carbon dioxide and oxygen in the air are determined by geochemical forces alone, life has nothing to do with it. They stated — and it's true — that the only net source of carbon dioxide is the output of volcanoes and the only net way of removing carbon dioxide is weathering — the degradation of calcium silicate type rocks. Living things, they said, process carbon dioxide and oxygen in a cycle that does nothing to affect their atmospheric abundances. Jim Walker and Dick Holland rejected both Gaia and Postgate on the grounds that they are not needed to explain the levels of carbon dioxide and oxygen in the air. So, you see these geochemists share in common with biologists like John Postgate a belief that their explanation of the regulation of the atmospheric gases alone is enough. Of course, contradictory criticisms like these from biology and geology, which cancel each other out, is no proof of Gaia's existence. They are just a sad consequence of a disastrously fragmented science. More seriously, there is a tendency in biology, possibly as a result of its long continuing battle with the religious fundamentalists, for biologists themselves to be dogmatic and argue like the Ayatollahs. They argue against any variation from what they see is the received text on neo-Darwinism, and incidentally a text that their saint, Charles Darwin, might well have denied. So strong is this tendency that Bishop Montefiore in his book *The Probability of God* takes quotations from eminent neo-Darwinists to illustrate that modern scientists are more dogmatic than modern theologians! I answer the criticism of the biologists and the geochemists not by trying to prove the existence of Gaia but rather by showing that, right or wrong, Gaia provides a new kind of thinking about the earth that is useful in itself. I shall now present a few examples of recent researches to illustrate just what I mean by this.

Cycles of nature

In two papers in the 1980s in *Nature*, Mike Whitfield, Andrew Watson and I first pointed out that the weathering of rocks and the deposition of limestone in the oceans are not just geochemical processes, like Jim Walker and Dick Holland had said, they involve organisms as well. Without life, the weathering of rocks — that is a digestion of rocks by the atmosphere — is much too slow to explain the current low level of carbon dioxide in the air. We suggested in our papers that the rates of rock weathering might be 30 times greater in the presence of life and that this was how Gaia might work. The suggestion at the time was speculative and based on the observation that there's a lot more carbon dioxide in the soil — something like ten to forty times as much — as there is in the air. However, recent measurements by two American scientists have shown that in fact the weathering of basalt rock, that is rock from volcanoes, by bacteria (let alone plants) is actually over a thousand times greater than is the weathering of the same rock under sterile conditions. This effect is so vast an increase in the rate of carbon dioxide removal that it's bound to have climatic consequences through the role of carbon dioxide as a greenhouse gas. It means that life will always try to keep the earth temperature low when the solar output is large, as it is now. The extent of climate control by this means can easily be calculated numerically in simple models using a computer. It suggests that regulation would take place at levels favourable for organisms, like now, and that the system will be very resilient to perturbations.

Up until the last two years, Gaian research was a very low key activity involving no more than about five scientists around the world. All of us worked on the topic in our spare time and none of us received any payment for doing it. In such circumstances, it was not practical to strive hard to develop tests for the existence, or otherwise, of global systems. That is a very expensive business indeed. So, it seemed to us to be better to go out in the world and collect information inspired by the predictions of the theory, whether it was right or wrong, and see where this led us. It didn't matter what we did as long as we stayed objective. A good example of this was an expedition I made in 1972 aboard the research ship *Shackleton* from the United Kingdom to Antarctica and back. The purpose of the expedition was to seek the presence of sulphur and iodine compounds in the ocean and in the air. You see, Gaia theory predicted that compounds of this kind are needed to transfer these elements from the sea, where they are abundant, to the land, where they tend to be scarce and made scarcer by their continuous removal as soluble salts dissolved in stream and river water. The measurements made during the voyage showed that the gases dimethylsulphide, methyliodide and others, were ubiquitous throughout the ocean environment. I should emphasise at this point that the very idea of the sulphur cycle including anything as exotic and bizarre as dimethylsulphide as a major component was entirely rejected by geochemists in the 1970s. They continued to reject it even after we published our papers following the voyage. Yet, from that voyage, part of the quest for Gaia, a new scientific investigation has grown and still grows. Indeed, recent discoveries suggest that this could be one of the most exciting areas of ocean and atmospheric science for today.

Let me tell you a bit more about the connection between the organisms in the sea, the algae, and this strange gas dimethylsulphide and the way it connects with clouds over the ocean and climate. At the end of the 1970s, a German scientist, Andi Andreae made a number of ocean expeditions to see if what I had reported was true and to try and find out the role of this strange gas. From these voyages he not only confirmed the measurements I'd made on the *Shackleton* but greatly extended them. He showed that dimethylsulphide was indeed the major sulphur compound of nature, the gas that really transfers sulphur around the world. In the early 1980s, Andi Andreae, two

American meteorologists, Robert Charlson and Stephen Warren, and I discussed the consequences of this strange gas in the sulphur cycle. In particular, we looked at the possibility that the atmospheric oxidation — that is the decomposition in the atmosphere of this dimethylsulphide — would produce droplets of sulphuric acid and other acids. If these droplets provided what the meteorologists called 'cloud condensation nuclei', the tiny droplets around which clouds form and without which them there would be no clouds. Instead there would be great blobs of water that fell out of the air as rain. Andi Andreae and I learned from the meteorologists that everyone was puzzled, because nobody had ever suggested a source of such cloud condensation nuclei over the oceans. The oxidation product of the gas we had found provided this missing link.

Previously, an Alaskan scientist, Glen Shaw, had proposed that the emission of other sulphur gases from biological marine systems would provide Gaia with a very cheap way of regulating climate and his notion was that these gases would oxidize in the stratosphere and there produce an aerosol of sulphuric acid droplets that would reflect sunlight and so cool the earth. This process would be cheap; a very little sulphur would have a large effect. The process might have been significant in some past epoch, but there was no evidence that it was playing any significant role at the moment. However, if the notion that these sulphuric and other acid droplets were the condensation nuclei of clouds was true then there would be a huge amplification of their effect. Light reflecting tendencies also would be amplified and it could become a large climate control mechanism. The four of us described our ideas in a paper in *Nature* in 1987.

The predictions we made linking cloud reflectivity to sulphur gas concentration and climate were tested by some satellite observations later that year by scientists from the National Centre of Atmospheric Research in the United States. They stated in their paper that our predictions were confirmed to within a few percent. It seemed as if we had discovered through a top-down Gaian approach a climate control mechanism as powerful as the greenhouse effect and potentially one that could oppose it and render untrue the gloomy predictions of unstoppable global warming. Although most atmospheric scientists have accepted this connection between cloud density and the gas I mentioned, the possibilities of self-regulation are still strongly resisted. In a *Nature* paper last year by Schwartz he argued from other satellite evidence that we were wrong. However, as you will be seeing in the next few months, we are supported by other independent scientists in papers soon to be published. This is a debate that is typical of science and the way the truth, ultimately, will come out.

Perhaps the most fascinating bit of evidence from the cloud-algae collection comes from glacier ice-core analyses taken in Antarctica. Those made by Professor Delmas of France show that, as one proceeds deeper into the glacial ice, the atmospheric carbon dioxide trapped in the ice remains constant until a depth corresponding to about ten thousand years ago is reached. At that point, it sharply declines to levels below 200 parts-per-million, during the times corresponding to the last Ice Age, and stays steady again at this low level throughout the period of glaciation. However, at the same time as the carbon dioxide declines in abundance, levels of a substance called methane sulphonic acid, which is a decomposition product of dimethylsulphide, rises to five times more than it was during the warm period that came after the Ice Age. Again, it was steady both afterwards and before the change. So, there was a possibility that during the Ice Age the earth was kept cool in part by a lower carbon dioxide and in part by a greater cloudiness above the oceans which reflected sunlight away into space. Perhaps I should add here that there are very sound reasons for believing that an Ice Age is a healthy stage of our planet, that Gaia likes it cold. It's unlikely that the full details of this story will be revealed for some years, but already it has become a topic of sufficient scientific interest to have justified the top-down approach to earth science.

A quite different yet complementary approach to measurements and experiments is to construct mathematical models and then see how well they explain the observed systems. You might think at first that such an approach is pointless. The feedback loops that link life and the environment are so numerous and intricate that you wouldn't have a dog's chance of making a model that would quantify or understand the whole earth. This is true of course, but during Christmas 1981, I made a simple model of the Earth. I made it to answer some biologists' criticisms who said that Gaia is a lot of rubbish because it implies there's purpose in the control of the earth. To have purpose by natural selection means you would have to have committees of the species meeting annually to decide what next year's climate would be. This is obviously ridiculous. It was a pretty good criticism and it set me thinking for a long time. It suddenly dawned on me over that Christmas that one could answer it by a simple model. The first of these models was known as Daisy World. I'm afraid for many of you it will be boringly well-known, but I will explain it again briefly, because although the system is remembered the details of it are all too often taken for granted and misunderstood. It is also the archetype of the more complex and more useful models of the physiology of the earth that have been developed and I won't be talking about in this lecture.

Daisy world

What I want you to do is to imagine a planet that is like the earth and is travelling at the earth's orbit around the star of the same mass and composition as our sun. This star, like all stars, warms up as it grows old. (I know astronomers are a pretty fickle lot, but it's one of the few things that they agree upon!) Anyway, this planet is spinning like the earth, but its atmosphere has few clouds and a constant concentration of greenhouse gases. In such circumstances the mean surface temperature can quite easily be calculated from the balance of the heat it's receiving from its star and the amount of heat it's radiating away into space. The colour, what the astronomers call the 'albedo', of the planet determines what its temperature will be. Now, I would like you to assume that this planet is well seeded with daisy seeds, some dark and some light. Growth rate is a simple parabolic function of temperature — the daisies don't grow if the temperature is below 5 degrees (it's too cold) and they don't grow if it's above 40. They grow best at about 20 degrees, somewhere in the middle. I also want you to assume that the planet is well watered and that there's no lack of nutrients. Now, if you assume all this, then it's quite easy to predict the area of the planet that will be covered by daisies from a knowledge of the mean surface temperature and the equations taken from population biology.

Conventional wisdom in science would see the evolution of that simple planet as follows: as the star warms up, the planet temperature rises. When the temperature reaches five degrees, the daisies — both dark and light — start to grow. At intermediate temperatures, they grow best. As the star continues to age, the planet temperature reaches 40 degrees, it's too hot and the daisies cease growing. That's the conventional wisdom. And I think it's a lot of nonsense.

This I think is what would really happen: as soon as the star warms up enough to bring the temperature of the planet to 5 degrees, daisies would start growing. In the first season, dark daisies would grow better than light coloured ones, quite simply because being dark they would absorb sunlight and be warmer. Remember, the planet is only just warm enough to grow daisies so the black ones would be the ones which would flower and leave seeds. There would be hardly any white daisy seeds. At the beginning of the next season, dark daisies would be off to a head start. Now, that's straightforward Darwinian natural selection, there are no tricks there. Dark daisies would grow and soon they would be warming not just themselves but the region in which they are growing. The temperature would start to rise with enormous positive feedback (biology sometimes produces positive as well as negative feedback it's worth remembering). As the regional temperature increases, into the intermediate temperatures, the population of dark daisies would increase also. Now, it wouldn't go on rising indefinitely, because once it got above the optimum temperature for daisy growth, growth would slow up. Not only that, but now it would be warm enough for the light coloured ones to grow and they would start competing by flowering and leaving seeds. Light daisies also reflect away some of the suns heat, giving it a competitive advantage at the higher temperatures, leading to a lessening of dark daisy numbers and bringing the temperature back down into the optimum range. And so, on it goes: dark to light to dark to light daisies. It turns out this simple competition between the two daisy species holds the planetary temperature very close to the optimum for daisies over a very wide range of solar luminosities. So there you have a system illustrating how Gaia can work without any foresight, planning or fudging.

I thought that this would convince my biological critics but it didn't. What they said was: 'Ah, yes, it's a very pretty model, but in the real world there would be cheats.' What do you mean by cheats? Well, daisies that wouldn't spend any energy making pigment and therefore would be at an advantage. They would grow and take over the whole system and it wouldn't work. Well, it's quite easy to put another daisy species in — it's just another differential equation. So, I put three daisy species in, one of which was grey and did nothing to the climate. I taxed the other two five percent of the growth rate for making pigment. What you get is a better regulation than with two daisies and of course any field ecologist will tell you that a more diverse ecosystem is better than almost any culture! Really, the cheats don't do very well. In fact, they don't do as well as either of the taxed light or dark ones. It's very obvious if you think about it. When the planet is very cold only dark daisies are fit to grow, nothing else can warm itself enough to take off. When the planet is very hot only light daisies are fit to grow, they're the only ones that can keep their cool. The only place that the cheats can grow are in the middle where regulation is not needed.

When I made this simple model, I didn't realize I was breaking all the rules of population biology. You cannot model more than two species simultaneously it's been said. If you attempt to do so the model goes chaotic, mathematics blows up. I was brought up in a school of science where you do the experiment and then read about it afterwards — luckily for me, otherwise I might have read the literature and never made the experiment! Blindly I went on and modelled the competition of ten different coloured daisy species, which is really overdoing it. However, what I discovered was a theoretical proof that diversity in the ecosystem, a wide variety of species, is a good thing. It brings more stability because when diversity is at a peak the system is regulating most efficiently. When it's regulating least efficiently, when it's just born or when it's about to die, that's when diversity is least. I think there's quite a message in that particular model.

Daisy World, as I've described it, is just an invention — a demonstration model to show you how I think Gaia works and why foresight and planning doesn't have to be evoked to explain automatic regulation. But, as the details of this kind of model are fleshed out, it becomes a generality and a theoretical basis for Gaia. I would like to think of it as the kind of model that the first population biologist, Alfred Lotka, who was a wise man and not a dogmatist, had in mind but could not develop in his day, 1925. At that time, there were no computers that could carry out the immense task of the hand calculation of even a simple daisy model of the kind I've just shown you.

Maintaining a healthy planet

Lastly, I want to tell you how Gaia and the top-down approach to the earth might help us in a better understanding of the problems for the earth, caused by the super-abundance of people and by their disturbance of the global environment. For example, I hope that the Green movement might learn something useful from this account of the top-down view of the earth, seen from the viewpoint of small science.

Environmental groups are often hostile to science and to scientists. I think they're right to be sceptical about the claims of big science, particularly when the hype and extravagance of some of the research is considered. There's an immense amount of self-seeking hype in that particular story, which has relatively little to do with either science or the well-being of people on the earth. We in science often forget that the science that the work we do reflects the needs and prejudices of the scientific community to which we belong. The scientific community is itself divided and very uncertain in the face of the world-wide decline in funding. Scientists are only human and for the most part concerned with their careers, their pensions, security and all that kind of thing. It isn't easy also for them to be responsible where there's no accountability, because part of the ethos of science is that you do what you like — nobody should have any right to tell you not to.

A consequence of all this is that the list of priorities often given by the Greens reflects a list not of what the Greens themselves think, but of the working priorities of the scientific community itself. They are not the priorities of the populace and still less the priorities of the planet. This is why until quite recently, I think, the Greens appeared to list global dangers in order of priority as follows: First of all, things nuclear — whether power stations, processing sites, waste disposal or bombs. Second, ozone depletion. Third, waste products of the chemical industry. In other words, the first three priorities are things that are carcinogenic and mutagenic to humans. However, they are also things that are very easy for scientists to measure and to work on. Interestingly, they are things that frighten us enough to make us quite ready to support, for example, cancer research agencies and the scientists who work on such topics. Now, I'm not for a moment suggesting that there's any kind of conspiracy to use fear for support for this kind of work. It isn't necessary. However, it took a character as strong as our own Prime Minister in her speech before the Royal Society last year to steal the flag from the Greens and stress the importance of planetary rather than human dangers. I think it's a bit of an indictment to all of us in the Green movement that this happened.

Before Margaret Thatcher's speech, the dangers from greenhouse gases — pollutants that threaten the planet like carbon dioxide and methane — would have been lower on the list, probably below acid rain, but still higher than the destruction of the forests of the humid tropics. Now, as an independent scientist with a top-down view, I see things very differently. To me, the vast, urgent and certain danger comes from the clearance of the tropical forests. Greenhouse gas accumulation may be an even greater danger in time to come, but not now. You see, the humid tropics are both a habitat for humans and a physiologically-significant ecosystem. That habitat is being removed at a quite ruthless pace, yet in the First World, at big prestigious scientific meetings, we try to justify the preservation of those tropical forests on the feeble grounds that they are the home of rare species of plants and animals — even plants containing drugs that could kill cancer. They may well do, they may well have these things, they may even be slightly useful in removing carbon dioxide from the air. However, they do infinitely more than this through their capacity to evaporate vast volumes of water vapour and gases and particles that assist the formation of clouds. Those forests serve to keep their regions cool and moist by wearing a sunshade of white reflecting clouds and bringing the rain that sustains them. Every year we burn away an area of forest equal to that of Britain and often replace it with crude cattle farms. Unlike farms here in the temperate regions, such farms rapidly become desert, so more trees are felled and the awful process of burning away of skin of the earth goes on. We don't seem to realize that once more than 70 to 80% of the tropical forest is destroyed, the remainder can no longer sustain its climate and the whole ecosystem then collapses. By the year 2000 at the present rate of clearance we shall have removed 65% of all of the forests of the humid tropics of the world. After that it will not be long before they vanish altogether and when they do it will leave about one billion poor living in those regions without support in what may well be a vast global desert. To my mind, this is a threat as great in scale as a major nuclear war, just imagine the human consequences, the suffering, the refugees, the guilt, and the political consequences of such an event. Moreover, it will happen at such a time when we in the First World are battling with the first surprises and disasters of the greenhouse effect, intensified by the extra heating caused by the clearance of those tropical forests. We may be in no position to help.

In addition to this impending catastrophe, if climatologists are correct in their predictions, and for example Gaia's effect is less than I thought, in the next few decades we are going to pass through one of the major transitions of our planet. We, our planet's first social intelligent species, are privileged to be both the cause as well as the spectators. The event I talk about, is, of course, a major climatic change. One that could be twice, or even six times, as great as the change from the last Ice Age until now.

During the last glaciation some tens of thousands of years ago, the glaciers reached the latitude of St Louis in America and the Alps in Europe. The sea level was at least four hundred feet lower than it is now and, as a consequence, an area of land as large as Africa was above water in the form of continental shelves exposed by the lower water level. This was covered by vegetation. You see what I mean by Gaia liking it cold — it was a lively and pleasant earth. The tropics were almost as warm as now, it was just that the poles were very much colder and larger. It was a pretty pleasant planet to live on and it was the home of simple natural humans who were just beginning to learn such neat ecocidal tricks as fire-drive hunting — set the forest on fire and a free effortless barbecue is available for all. We were just as bad then as we are now!

To understand what is before us in the next century, imagine a changing climate at least twice as large as that of the Ice Age until now. That in truth, is really what the climatologists are talking about: the start of a Heat Age. The temperature and sea level will decline decade by decade until eventually the world will become torrid, ice free and all but unrecognizable. 'Eventually' is of course a long time ahead and it might never happen. What we do have to prepare for now are the events of the transition itself. Events that are just about to begin and these are likely to be surprises as unpredictable to big science models as was the ozone hole itself. There may be extremes, like storms of great ferocity, or unexpected atmospheric events. You see, nature is very non-linear and unpredictable and never more so than in a period of transition. But what of Gaia? Will she not respond and keep the status quo? Well before we expect Gaia to act, we should realize that the present inter-glacial period as I have mentioned could be regarded as a 'fever' for Gaia. Left to herself, she would be relaxing into her normal comfortable and healthy state of an Ice Age. She may be unable to do this relaxation because we have been busy removing her skin and using it as farmland, especially those trees of the forests which otherwise are amongst the means for her recovery. Also, we are adding a vast blanket of greenhouse gases to an already feverish patient. In these circumstances Gaia is more likely perhaps to shudder, to move over to a new stable state fit for a different and more amenable life species. It could be much hotter, it could be much colder, but whatever it is it will no longer be the comfortable world that we now know.

The only ray of hope I have is the discovery that came from thinking about Gaia, the discovery of the dimethylsulphide gas and its connection with the algae living over the oceans and the clouds above them. Without this gas there would be fewer and less dense clouds over the oceans of the world and without the white clouds over the oceans the earth would be much hotter, for the dark oceans absorbs the sun's heat but the white clouds reflect it back to space. So, if the algae grow more and give off more of their gas as the world warms up, then to some extent the heating of the gaseous greenhouse might be ameliorated. The human and political consequences of the two geocidal acts — forest clearance and suffocation by greenhouse gas — will be the news, news that will usurp the political agenda. Soon and suddenly in the humid tropics there could be a billion or more humans suffering drought and floods, perhaps with mean temperatures like those of the Australian deserts. They would be without support in that vast arid region when we could no longer help them. These predictions are not fictional doom scenario, but uncomfortably close to certainty. I think we are like a modern version of the Gadarene swine driving our polluting cars heedlessly down the slope into a sea that is rising to drown us.

I have spoken as the lone independent scientist, but I'm far from being outside human concerns. I have eight grandchildren and wish to see them grow up in world that has a future for them. Nevertheless, following an independent Schumacher-kind of lifestyle has thrust me to the vacant position as the representative, the shop steward to the rest of life on this planet. My constituency is all life other than human and includes the bacteria, the worms and the less attractive forms of life. I have to do this because there are so many who speak for people, but so few that speak for these others on whom the planet depends — much more than it does on us. From my laboratory in west Devon, I can see the stars at night in the Milky Way and in the daytime I can hear the birds and smell the earth. To see and feel the earth in this way, and to think of it as a living organism, gives substance to that Christian concept of stewardship and turns our hearts and minds to what should be our prime environmental concern: the care and protection of the earth itself and especially the forests of the humid tropics. It's not enough to be concerned for people, there is no tenure for anyone on this planet, not even for a species. If we do not recognize our responsibility to our planet we may not as a species even reach our allotted span. So let us be moderate in our ways and aim for a world that is healthy and beautiful and which will remain fit for our grandchildren as well as those of our partners in Gaia.

Complexity, Creativity and Society

Brian Goodwin

Professor Brian Goodwin was born in Montreal, Canada, studied biology at McGill University, mathematics at Oxford, and took his Ph.D. in embryology at Edinburgh University under the eminent biologist C.H. Waddington. After research appointments in Canada and the U.S.A. he returned to Britain and took up a position as Reader in Development Biology at the University of Sussex, and in 1985 was appointed Professor of Biology at the Open University. He is now emeritus professor. His most recent book, How the Leopard Changed its Spots, *was awarded the Scientific and Medical Network Book Prize for 1995. This lecture was delivered at the 1989 Mystics and Scientists conference.*

Every once in a while the dialectic of science carries ideas in unexpected directions that connect with wider social movements, revealing the deeper relationships between the two. Current concerns about health, environment, community structure and quality of life in general as opposed to quantity of consumer goods, all reflect shifts of focus from the individual to relationships and the collective, the part to the whole, and from control of quantity to participation in quality. This article is about new ideas in science that are connected with the recent remarkable proliferation of books and articles on chaos and fractals and their relevance in diverse areas of science and society. The broader context of these developments is the study of complex systems, which we all know are the reality which we experience, rather than the simple models of reductionist science. So, in a sense, science is finally coming of age and as it does so it is going through some fundamental soul-searching. There is no consensus yet about where we shall end up, except that it will definitely be somewhere else. What follows is my own assessment of the significance of the new developments for the future of science and some of its social consequences.

The discovery of strange attractors

Complexity has a multitude of colloquial meanings, none of which corresponds precisely to its use in the area of study that has come to be known as the science of complexity. There it refers to the potential for emergent order in complex and unpredictable phenomena. This science grew out of puzzling problems concerning planetary motion that were uncovered by the great nineteenth century mathematician and physicist, Henri Poincaré. He noticed that something as apparently simple as three bodies interacting, such as sun, earth and moon, give rise to very strange dynamic motion that appeared to carry a distinct signature, a pattern that had new and unforeseen properties. Working out the precise characteristics of that signature has occupied mathematicians for nearly a century.

However, the study of a rather arcane problem in planetary dynamics is not itself sufficient to bring mathematics to the attention of millions: the maths has to connect with human experience. This connection was made when Edward Lorentz, a meteorologist at the Massachusetts Institute of Technology, discovered the same complex behaviour as Poincaré had while studying, in the 1970s, the solutions to equations describing weather patterns. The difference was that Lorentz had a computer, so that he could get it to trace out the solutions on a screen. Using Poincaré's method of mapping complex dynamic patterns, he observed a new and beautiful mathematical object, now known as the Lorentz attractor (Figure 1).

Lorentz realized that he was dealing with a radically new type of behaviour pattern whose properties led him to an immediately graspable metaphor: a butterfly flapping its wings in Iowa could lead, via the strange dynamics of the weather, to a typhoon in Indonesia. Stated in another way: very small changes in initial conditions in the weather system can lead to unpredictable consequences, even though everything in the system is causally connected in a perfectly deterministic way. The way this works in relation to the figure is as follows. Suppose you choose any point on the tangled curve in Figure 1 as the starting point, corresponding to some state of the weather. This will develop in a perfectly well-defined, though complex, manner, by

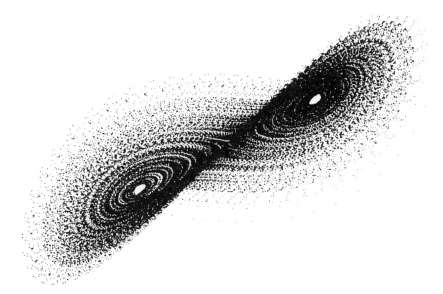

Figure 1. A Lorentz attractor, discovered while modelling weather patterns.

following the curve from the (arbitrary) starting point in one direction, which is prescribed by the equations. Every successive state is clearly defined — that is, everything is perfectly deterministic, since this is what dynamical equations describe. However, suppose there is a small disturbance that shifts the weather to a neighbouring part of the system, to a point on a nearby part of the tangled curve. Then, comparing the state of the initial weather system with that of the disturbed system as they both develop along the curve, a basic property of the strange attractor is that they move away from each other exponentially fast. That is, knowing what the weather is now is no predictor of what it will be a couple of days hence, because tiny disturbances (the butterfly effect) can produce exponentially divergent behaviour. This is the signature of deterministic chaos, now identified in a great diversity of mathematical equations whose dynamic properties are described by strange attractors, of which Lorentz's attractor is an example.

The consequences of this mathematical discovery are enormous. Since most natural processes are at least as complex as the weather, the world is fundamentally unpredictable in the sense that small changes can lead to unforeseeable results. This means the end of scientific certainty, which is a property of 'simple' systems (the ones we use for most of our artefacts such as electric lights, motors, and electronic devices, including computers). Real systems, and particularly living ones such as organisms, ecological systems and societies, are radically unpredictable in their behaviour, as we all know. Long-term prediction and control, the hallmarks of the science of modernity, are no longer possible in complex systems and an appropriate relationship to them is to participate with sensitivity in their unfolding, since we are usually involved in their dynamics in some way. Insight into the processes involved is now assisted by a precise hypothesis about what may underlie this complexity: they may all live dynamically on strange attractors, obeying dynamic rules that lead not to stereotyped but to unexpected behaviour.

This is where another aspect of the complexity story enters. A typhoon may well be the unforeseen consequence of the butterfly innocently seeking nectar in the fields of Iowa. But a typhoon is not itself a chaotic weather pattern: it has a highly organized dynamic structure. That is to say, a typhoon is one of the possible patterns that emerges from the complex behaviour of the weather system. So the dynamics of the weather combines both order and chaos. They live together. Although we cannot predict what will be the consequences of a small disturbance, we do know that one of a limited set of possibilities will follow — a typhoon, a high pressure region with sunny skies, a low pressure front with rain, and so on — a large but not indefinite set of possible emergent patterns. This is the signature of complexity.

Complexity and the edge of chaos

This approach to understanding complex processes is now being applied to many areas of study: to evolution, for instance, to economics, and to the dynamics of social change. All of these combine unpredictability with order in distinctive ways. Organisms of different species — kestrels, badgers, oak trees, columbines — all have distinct properties that express a specific intrinsic order that we use to recognize them as a 'type,' a species of organism. However, both the emergence of these species in the course of evolution and their extinction are fundamentally unpredictable events. Developing models of these processes is the concern of 'complexologists' interested in biological phenomena.

The same approach can be used to study physiological activities within organisms, looking at the dynamics of the immune system, for example, or of the heart. Medical cardiology is undergoing something of a revolution as a result of applying ideas from complexity theory to the study of normal and abnormal heart beat patterns. Nothing is more orderly than the rhythmic beating of your heart as you sit reading this, you might think. It is the paradigm of physiological regularity on which your life depends in a most immediate way. However, combined with this order there is a subtle but apparently fundamental irregularity: in healthy individuals, and particularly in young children, the interval between heart beats varies in a disorderly and unpredictable way. If the interbeat interval is regular — either constant or itself rhythmic — then this is a sign of danger, indicating that the heart may be prone to fibrillation leading to cardiac arrest, or some other arrhythmia. Cardiologists such as Ary Goldberger, working at the Beth Israel Hospital in Boston, consider that the healthy heart beat is embedded in a sea of chaos which both reflects its sensitivity to the activities of the rest of the body and its capacity to respond appropriately to the continuously varying demands made upon it as we go about our diverse activities and experience different emotions. Too much order in heart dynamics is an indicator of insensitivity and inflexibility, just as rigidity in other patterns of behaviour indicates stress and incipient danger. An intuitively attractive idea follows: complex adaptive systems function best when they combine order and chaos in appropriate measure.

This idea was in fact one of the earliest suggestions to come from the analysis of complex systems. While studying the dynamic behaviour of cellular automata, which are particularly useful for modelling complex systems, Norman Packard and Chris Langton had the insight that the 'best' place for these systems to be so that they can respond appropriately to a constantly changing world is at 'the edge of chaos.' Here order and disorder are combined in such a way that the system can readily dissolve inappropriate order and discover patterns that are appropriate to changing circumstances. This fertile suggestion has been subject to severe criticism, as should any proposal that attempts to capture a generic property of a whole new class of systems. However, the basic idea that creative, adaptive systems are most likely to function best at the edge of chaos is proving to be a robust insight, despite the difficulty of pinning it down precisely (mathematically, logically). One of the foremost contributors to the elaboration of this concept in a variety of contexts, all of which relate to basic biological processes such as the origin of life, gene activity patterns in organisms, and the evolution of species, is Stuart Kauffman.[1]

From competition and survival to creativity

There is a very interesting paradox about evolution that emerges from this approach, which has significant social resonances. In the Darwinian perspective, what drives evolution is competition for scarce resources between organisms that differ from one another in their 'fitness,' their capacity to leave offspring. The survivors of this struggle are the better adapted, those that can function better in their environment. However, the evidence from studies of species emergence and extinction during past geological ages and from models that simulate these processes, is that species do not go extinct because of failure to adapt to changing circumstances, or because of cataclysmic events such a meteorite impacts or volcanic eruptions. Although these have undoubtedly contributed to the disappearance of species such as the dinosaurs, for example, it appears that there is an intrinsic chaotic component to complex systems such as interacting species in ecosystems which results in intermittent extinctions that vary from small to large, with a characteristic distribution, and these occur independently of the sizes of external perturbations. As David Raup put it: species go extinct not because of bad genes but because of bad luck.[2] There seems to be an unpredictable dynamic to creative processes such as evolution that involves inevitable extinction or destruction with a characteristic pattern of survival that is not due to individual success or failure but to the interactive structure of complex processes. The game of life, we might say, is one of creative emergence and extinction in which the reward is not long-term survival but simply transient expression of a coherent form, a revelation of a possible state of life which we call a species, whose value is intrinsic to its being.

Clearly the metaphors are shifting here from competition and survival to creative emergence and expression of appropriate novelty. These are not necessarily in conflict. They express different ways of seeing complex processes, the latter viewing the dynamics of the whole while the former expresses the perspective of a part. In fact, the science of complexity has been characterized as a holistic science which seeks to describe the properties of complex wholes. For example, whereas a knowledge of parts and their malfunction gives the medical model of disease, a study of the dynamics of the heart described earlier provides an example of how one can assess the health of the whole body from a study of the behaviour of a part, the pattern of the heart beat.

Diagnosing the state of the whole from a study of a part is of course the procedure used in a variety of complementary therapies, such as pulse diagnosis for acupuncture treatment, reading and treating the body via the feet as in reflexology, and so on. It is the traditional therapeutic approach. The science of complexity may provide useful diagnostic procedures based on a similar conceptual model in which the relationship between whole and part is so intimate and integrated that the dynamics of the part reflects the condition of the whole body. However, the concept of a whole within the science of complexity is something new and involves an apparent contradiction. Because of the mixture of order and chaos that is involved within a complex whole, the parts simultaneously have significant freedom while expressing order. Put metaphorically, the next heart beat can occur whenever it likes (i.e., whenever the heart 'decides' is the appropriate moment within the current situation that it experiences within the body, and in terms of its own complex dynamics). The only constraint is that there be an overall average rate of beating for any particular physiological state (sleeping, sitting, running, together with the emotions experienced). More precise statements of this paradoxical condition of maximum freedom of the parts together with coherence of the whole can be given in terms of entropy and concepts derived from the quantum mechanical definition of coherent states.[3] This holistic view has developed independently in a variety of contexts, including quantum mechanics itself.[4]

Creativity and society

These concepts are now being applied to a range of complex processes, from ecosystems to social organizations. Business corporations have been among the first to see the potential relevance of these ideas to management structure and creative organizational change. Since their everyday experience is 'living on the edge,' any insights into dynamic structures that facilitate adaptive response are welcomed. The suggestions of complexity theory for business practice are a flattening of the management hierarchy, distribution of control throughout the system with fluid networks of interaction between the parts, and the necessity of periods of chaos for the emergence of appropriate new order. The move towards a more anarchic, spontaneous dynamic is clearly threatening to the controlling manager, but it appears to be the path to creativity and diversification. This in no way guarantees survival, just as there is no long-term survival granted to adapted, adapting species in evolution. What it allows for is innovative expression, which has intrinsic value for the members of the enterprise as well as providing the best chance of the organization persisting in a constantly changing corporate world. All the participants in this sector of social organization can then experience a higher quality of life since they have greater freedom, more opportunities for creative play, and richer interactions — good for them and good for the organization. The primary goal would not then be survival through maximization of profits, but a fuller and more creative life for all members of the company and in so doing to maximizing the chances of appropriate collective response to perpetually changing circumstances. The shift of focus here is towards quality and away from quantity as the goal of a more mature society. In fact, the science of complexity may be seen as a significant harbinger of change from the dominant science of quantities that has characterized modernity towards a science of qualities that is emerging in the post-modern era.[5]

Naturalistic ethics

There are clearly many implications of this shift of perspective in science that concern our understanding of the natural world and our relations to it. One of the most profound concerns the relationship between scientific knowledge and ethical action. Logical positivism, the philosophical basis of empirical science, completely severs any connection between 'facts' and 'values' by asserting that no human mental components should contaminate the pure observation of reality by the senses. The outrage felt by some philosophers of science at the corrosive effects of this stance is clearly expressed by Rom Harré:

> Progress in the philosophy of science, as thus conceived, came to an abrupt halt in the twentieth century, with the rise to dominance in the academic community of a corrupting and

deeply immoral doctrine, logical positivism. It was positivistic
in that it restricted the content, source and test of scientific
knowledge to the immediate deliverance of the senses. It was
logicist in that it confined the task of philosophers to the
laying bare of the logical form of finished scientific discourse.
The immoral character of this viewpoint hardly needs spelling
out these days. It is highlighted, for example, in Habermas's
protest against the importation into human management of
those favourite concepts of the positivist point of view,
'prediction' and 'control.' It is the animating philosophy of the
morally dubious authoritarianism of Skinnerian psychology.[6]

Harré's philosophy of science is a form of realism in which
explanation of phenomena depends upon the construction of models or
theories that go beneath the level of appearance to the generative processes that describe the real underlying causes of phenomena. An
example of this is the theory of fluids (liquids, gases) that is used to
describe the phenomena of the weather. Such theories use the fundamental concept of a field, a domain of relational order in which the
state of one part of a fluid is connected by a precise mathematical
relation to the states of neighbouring parts so that the whole weather
system consists of a single three dimensional field which unfolds in
time according to particular laws. Poincaré's original studies on
planetary dynamics involved the use of gravitational field theory to
describe the motions of three interacting bodies as an integrated whole.
    This is the basic stuff of science, and Harré is actually rescuing its
philosophy from an aberration. But he has added some distinctive
components, in particular a rethinking of causality in science.[7] This
allows us to escape from the curious impasse imposed by David Hume
in which we are unable to perceive necessary causal connections in
natural processes and can do no more than register correlations
between causes and effects, which at any moment could change. This
is the ultimate fragmentation of atomism and the denial of our ability
to participate in and grasp natural processes. Harré and Madden restore
the notion of causal powers to nature, which is then reinvested with
'powerful particulars' that underlie the appearance of specific structures such as the motion of the planets, weather patterns, and the
strange attractors of chaotic dynamics. These powerful particulars are
the real underlying causal powers of natural phenomena, particular to

each type of field such as the force of gravitational attraction, of electromagnetism, or of fluid dynamics.

With generative processes comes the concept of natural kinds, those structures that we identify as having distinctive types of intrinsic order such as the elements in physics (carbon, copper, gold, etc), typhoons and other weather patterns, and species of organism (kestrels, badgers, oak trees, human beings, etc). This argument is developed with clarity and force by Roy Bhaskar in books that develop the realist philosophy of science in both the natural and the social sciences.[8] With natural kinds comes a naturalistic ethics. This depends on our relationship to what we take to be the truth, in so far as science reveals it to us. If we believe that we have some insight into the true nature of something, then this influences our behaviour towards it. For example, if we believe from the evidence that certain foods are beneficial to human health and others are detrimental, and we are in a position to influence diet, then we will experience some compulsion to encourage the consumption of healthy rather than unhealthy foods. If we believe that children have a basic need to play (i.e., this is a necessary expression of their nature), then in general we allow space and opportunity for them to play. Likewise if, from close examination and study of the behaviour of chimpanzees, particularly in the wild, we become convinced that to express their natures they need conditions such as life in social groups and an environment with adequate freedom of movement, then we will discourage the confinement of chimps in isolation or in severely restricted circumstances. The same type of argument holds for whatever understanding we may have of any other species.

Andrew Collier, discussing these issues in an admirably clear book, puts the argument for the social sciences as follows:

> Social sciences, then, generate practical emancipatory projects by showing there to be (a) a need, (b) some obstacle preventing its satisfaction, and (c) some means of removing this obstacle. This is not a matter of mere technical imperatives, coming into play only if you want the projected good; given that a social science can tell us not only about the means of satisfaction but also about the need itself, it may ground assertoric imperatives, that is, since you need this, remove that obstacle thus.[9]

Furthermore, according to Bhaskar: 'for emancipation to be possible, knowable emergent laws must operate,' thus linking us back firmly into scientific realism.[10] This is necessary because without principles of emergent order from complex systems there is no real nature that is expressed and nothing intelligible for us to understand, everything being totally accidental and contingent. Then all possibilities would have equal logical status and we could rationally argue that there is no violation involved in depriving children of play and requiring them to work; or that we can design chickens and cattle to have extra muscle that prevents them from moving 'naturally,' since we can also design cages that can support them. The realist argument also connects directly with the programme of research in the sciences of complexity, in which a primary goal is to understand the generative processes that give rise to the diverse forms of emergent order that arise in evolution (biological and cultural).

It thus becomes clear that the move from control and manipulation of nature to participation in and understanding of the creative expression of nature in emergent forms has significant consequences in a variety of contexts. Among these are our relationships with the environment and our use of plants and animals as commodities for manipulation by genetic engineering and biotechnology. The realist argument that different types of organism have natures, developed in a book by Webster and Goodwin, means that, in so far as we understand these natures we are in a position to remove obstacles to their expression.[11] Only when organisms are able to express their specific natures will they be healthy. If we are to consume these organisms (lettuces, cabbages, fish, and so on) as food, then food quality depends upon the health of the organisms and hence on the conditions under which they live.

We are currently manipulating species more and more, both through changes in their environments (soil conditions, use of fertilizers, herbicides, pesticides, and so on) and their genes in order to increase quantity (of product and of money). In so doing we are disturbing complex systems in unpredictable ways that result in epidemics such as BSE in cattle and innumerable food allergies. We are pushing species to the brink of their stability, where they suffer disease such as increased mastitis in cattle treated with growth hormone to increase milk yields. What is so absurd about this example is that milk production was adequate before the new technology was introduced, so it was quite unnecessary. Quantity goes up, quality goes down, and these complex living systems are driven beyond their intrinsic stability (their natures) into states of propagating instability that manifest as epidemics and disease.[12] This is driven by intellectual property rights and short-term profit. Frequently there are alternatives that are ecologically safe and maintain health, though not patentable and therefore not attractive to the food industry. Biotechnology has a role to play in food and drug production. But companies need to act responsibly, with adequate biosafety controls and regulations. These may well require periods of testing extending beyond ten years because of the slow response times of complex genetic networks and gene transfers between species in ecosystems, which need to be studied intensively under controlled conditions before transgenic varieties are released. Without these, human health and that of the ecological networks on which we all depend will be put continuously at risk, with periodic devastating epidemics. In the past, only litigation after severe damage has forced responsible behaviour. The more sophisticated understanding of complex systems that is now emerging encourages us to take precautionary measures to avoid such disasters, reflecting sensitive participation in the processes in which we are immersed rather than continuing with the fiction that we can exert control from the outside as objective, detached observers with predictive knowledge of outcomes of manipulations.

The potential influence of a new science that invites participatory engagement in complex systems, whether physical, biological, social, corporate, or economic, is very great. I have touched on only a few of the issues that are coming within the purview of new developments that could link science more closely, and more responsibly, to major social issues.

References

1. Stuart Kauffman, *At Home in the Universe,* Oxford University Press, Oxford 1995.
2. David Raup, *Extinction: Bad Genes or Bad Luck?* Norton, New York 1991.
3. Mae-Wan Ho, *The Rainbow and the Worm,* World Scientific, Singapore 1993.
4. David Bohm and Basil Hiley, *The Undivided Universe,* Routledge, London 1993.
5. Brian Goodwin, *How the Leopard Changed Its Spots,* Weidenfeld and Nicolson, London 1994.

6. Rom Harré, *Varieties of Realism,* Basil Blackwell, Oxford 1986, p. 21.

7. Rom Harré and Edward Madden, *Causal Powers,* Basil Blackwell, Oxford 1976.

8. Roy Bhaskar, *A Realist Philosophy of Science,* 2nd edition Harvester Press, Brighton 1978; also *Scientific Realism and Human Emancipation,* Verso, London 1986.

9. Andrew Collier, *Critical Realism,* Verso, London 1994.

10. Roy Bhaskar, 1986 (see note 8).

11. Gerry Webster and Brian Goodwin, *Form and Transformation,* Cambridge University Press, Cambridge 1996.

12. Mae-Wan Ho, 'Unravelling Gene Biotechnology,' in *Soundings* 1, pp. 77–98.

The Biology of Being:
A Natural History of Consciousness

Lyall Watson

Dr Lyall Watson studied medicine at the University of Witwatersrand, South Africa, then switched to biology and took a Master's degree at the University of Natal. He took a Ph.D. in zoology at London University writing his thesis on animal behaviour whilst working under Desmond Morris at the London Zoo. He then studied anthropology at the American School of Oriental Research, Jerusalem and archaeology at Petra in Horan and Kano in Northern Nigeria. From '65 to '67 he was Director of the Johannesburg Zoo and afterwards returned to London to join the Science Features Dept. of BBC Television. Subsequently he founded a Life Science consultancy for film making and expedition work and led expeditions to the Indian Ocean, Indonesia, Antarctica and the Amazon. His published books included Gifts of Unknown Things, The Romeo Error, Omnivore *and* Super-nature *which was a worldwide bestseller; also* Beyond Supernature *and* The Nature of Things. *This lecture was delivered at the 1988 Mystics and Scientists conference.*

A few years ago, I found myself, as I sometimes do, in flight from a world that threatened to file my facts, process my words and daisy-wheel me to death. I agreed in a moment of madness to deliver a friend's yacht to the Seychelles. This relatively simple task, a voyage under sail across an ocean I know rather well, became somewhat complicated when the old wooden hull of the boat sprang a leak and began to sink beneath me. I turned downwind and headed as fast as I could towards the nearest shore. Three heartstopping days later, I made a landfall on a deserted coast. I nursed the boat into the shelter of a protected estuary, beached it gently on a sandbar and tied it fore and aft to a pair of convenient palms. I went ashore in search of carpentry a little more skilled than my own. The beach on which I landed was familiar enough, the mix of the usual mangroves and tropical hardy shrubs, which you find on most shores in the equatorial area. However, behind the first line of sand dunes I came across a landscape which left me absolutely breathless.

I had never seen the like. There were plants there to be sure, but they were grey rather than green, and they seemed at first sight to consist of nothing but a welter of tendrils that sprung up from the red earth into the sky — like little vegetable funnels, miniature tornadoes, tentacles from a million giant and demented squid. Some of them were thick-limbed like sausages which collapsed under their own weight. A few of the thinner ones swarmed in a serpentine fashion across the ground, raising every now and then a tall and menacing quill. Every one of them, regardless of their size and dimension, was covered in a formidable array of prickly, fiendish spines. The whole effect was other-worldly, like looking at a set for a science fiction film, something that might have been designed by George Lucas.

The effect was heightened by what I could see moving about in the vegetable nightmare. There were birds there, but birds like none I had ever even heard of. They looked wrong. They looked as if they'd been put together by a five-year old with a do-it-yourself bird kit and a vivid imagination. There were butterflies, some as big as pigeons, and they drifted along like disembodied birds. There were beetles of inconceivable colours, patterns, forms, which I couldn't put into their nearest family, let alone give a proper species name. Now that may not sound upsetting to you, but believe me, for a field naturalist of my experience, it was downright embarrassing! I found myself in the biological equivalent of culture shock. I felt like a traveller who has strayed off the edge of one of those early maps that leave large blank spaces labelled with the warning 'here be dragons.' It was as though I stood in a strange country, unable not only to understand the language or read the signs, but even to be sure that the inhabitants were human at all. It was as though I had landed on another planet.

A forgotten world

My landfall, in fact, was on the southern coast of Madagascar and I was looking at its spiny forest. It is forest that has evolved on that lost world during the 100 million years that have passed since it sailed away from Africa and drifted off into the Indian Ocean with a cargo of living fossils. Ninety-eight percent of the species of Madagascar are found nowhere else in the world, and travelling there today is like catching a time machine. It's like landing on another world in another dimension. I know of no other place so foreign, so exotic, so alien. At that time, there were are no text books, no field guides, nothing to help. For a biologist it was both exciting and disturbing. It provided a sort of ambivalent *frisson* I think the astronauts must have felt when they first walked on the face of the moon. It's an encounter with otherness. And such an encounter, I have found, can be a very useful thing.

Every time I go to Madagascar I am liberated, I can start afresh. Once I get over the shock of not knowing, once I can accept that it's alright not to know, that *nobody* knows, then it's like being born again, like being a child once more. Someone who is able to see and experience things from scratch without preconceptions, to know things as they are, or as close to that reality as we are ever likely to get. It's a Garden of Eden experience, a return to innocence, but a return with the advantage of a new awareness, one that only can be produced by having already eaten of that tree of knowledge. One comes to this garden with some basis for comparison and comprehension. I still can't put names to many of the things I see there — the snakes are bizarre! — but I relish the luxury in Madagascar of knowing that not one of those snakes is poisonous. They are all civilians, the clerks and secretaries, of the serpent world, isolated there before venom was invented. So, it's a Garden of Eden in which the serpent is toothless, a place in which one can relax and yet take nothing for granted, for you don't know what is going on. It's a perfect place, in other words, to think about consciousness, to enjoy the luxury of letting go and looking inside the system and inside yourself, to wonder about the biology of being, of what you are, what we all are.

When I went to back to Madagascar at the end of last year, I took the chance to walk again in that spiny forest and to think about creation, and about consciousness. I have written about consciousness several times before and I've talked about it a lot. However, I say what I have to say in the full knowledge that consciousness is a subject that comes with its own peculiar Catch 22: If the human brain were so simple that we could understand it, we would be so simple that we couldn't. There is no getting around that. There is no avoiding the additional complication, introduced by Wittgenstein, who pointed out that all we see when we look inside ourselves is what amounts to a beetle in a box — a beetle only we can see, a beetle we can never compare with anybody else's beetle. Now, that too is an undeniable fact and it is a sobering thought in Madagascar where the chances are that every beetle you see is new to science.

The spooky thing about Madagascar is that everything from that strange island defies public definition. Every view, every discovery has the quality of an inner experience, something difficult or perhaps impossible to discuss, or to describe out loud. Yet, each time I come back from Madagascar that's exactly what I try to do; something I feel the need to do, because the experience there has been important to me. I imagine, too, it ought to be interesting to other people, so I keep trying. Despite all the shortcomings of language and in spite of the logical difficulties involved in trying to convey a sense of what it's like to be in a place like Madagascar, this sort of apparently hopeless cause is one which humans everywhere keep on espousing. We persist in trying to communicate the ineffable, we keep on trying to do it through a medium of language, with a vocabulary which just isn't designed to cope with that sort of problem. We may never get a glimpse of each other's beetles, but every culture has a rough set of rules for describing such things. The rules may be arbitrary but they are communal, they are publicly agreed and they are written large on the lid of everybody's beetle box.

People always agree that within their bodies there exists a spirit, an entity conscious of its own existence and of its continuity in time, a creature that feels and perceives and wants. We all suspect that somewhere inside there is an I, an inner self. All people of all the human cultures we know about agree on this; there is my body and then there is me. It is the I that wills that body to move, that makes patterns of incoming sensations, that knows passion, that feels pleasure. There is universal agreement about such a dichotomy and not a shred of evidence to prove that it exists. There is nothing to explain the fantasy except a certainty that it is so. We know the people we meet do indeed have beetles in their boxes and aren't telling us lies about completely empty containers. I've always been disposed to take what people say seriously. I'm a fan of folklore. I'm a supporter of superstition. I'm loud in my praise of old wives and their tales. I'm strong in my admiration of childish perceptions. I'm drawn to the truth one finds in common sense and common understanding. In Madagascar, for instance, I share my awe, my astonishment, of the natural history there with other strangers — ten million of them. That is the whole population of Madagascar, because all people are new to Madagascar.

The island was untouched by humans until 1500 years ago when the first immigrants arrived from Asia and Africa, and they still haven't quite settled in. They give the impression if you talk to them, if you live with them for any length of time, of living on a frontier, of keeping their spiritual suitcases packed, ready to move again. However, they do what they can in the meantime to restore an equilibrium, which they are vaguely conscious that they themselves have disturbed. The problem is that the herding of cattle and the cultivation of rice, neither of which belong in Madagascar, are destroying the island. They are wiping out its forests so fast that a quarter of a million indigenous species have already gone. Yet, the strange thing is that these foreign people, these foreign farmers, have somehow found an affinity with the local flora which, given its incredible novelty, is utterly unlike anything that occurs anywhere else. They cannot have arrived at this affinity in the few centuries that have passed since they first arrived, and they can't have done it on the basis of trial and error. It just isn't possible, there hasn't been time. Despite that, native herbalists in Madagascar have nevertheless built up a very impressive pharmacopoeia of useful plant remedies, which they use in their everyday lives. The question about this, which intrigues me, is what led them to select some plants, or some parts of some plants, for particular purposes and reject the hundreds of thousands of other species.

I asked the medicine man who lives near the spiny forest this very question. I asked him what it was that led him to certain plants rather than others. He was quite clear about it. He said that the secrets were given to the people by the plants themselves. He wandered around when he needed a remedy with his eyes almost closed. In that frame of mind he was led to certain plants and certain flowers and certain shrubs by what he said was their sound. He took and tested these by holding a petal or leaf in the palm of his hand, or placing it on his tongue. When he did that, he could feel the effect of that particular plant on the appropriate part of his body. Some felt cold, some felt warm and the best were accompanied by a feeling so strong that it made his head swim. A few which gave him pain or made him vomit he discarded in a hurry, but a large number of other plants, despite their unprepossessing appearance, were friendly, were open to communication. I am beginning to suspect that, despite everything I learned as an academic botanist, that such interaction is really possible, that there is an exchange of information at a very basic level between all living things.

There is a flow of pattern, of instruction, which allows even radically different organisms, as different as a man and a bush, to borrow each other's ideas. Some animals, for instance, seem to know what is good for them without being taught. You know of animals which travel great distances to find salt, of sick sheep and even domestic dogs that seek out and eat certain grasses which have a purgative quality. I know of a record of a mountain goat with an injured leg that actually made itself a poultice of lichen and clay and applied it to the wound. I saw a case in South Africa recently in which cattle were eating the bark from a tree they'd never bothered to eat before. When the tree was examined, there was embedded in the trunk a single copper nail. The particular problem that was plaguing the cattle at the time was a deficiency disease caused by lack of that particular trace element. I've seen lemurs in Madagascar pick the leaves from certain plants and bury them in shallow pits in the forest floor where they ferment, in a fermentation pit technique which many people still use. They then come back two or three weeks later and dig up the products of that fermentation, which after analysis we now know contains a very powerful stimulant. Lemurs have always known this. The list can go on. I've given you a few, just enough to give you a hint of where I'm heading.

The very strangeness of being in Madagascar encourages one to think in new ways. The lack of familiar signposts, of landmarks, makes it possible to look at a new ecology in a new light, without the labels one is accustomed to putting on things. I have to admit that there are some demarcation lines. People are different, not only from each other, but also from other living things. However, perhaps this is true only for those things that lack recognizable personalities.

Ants and bees, to my mind, are our social equals in many ways. They are so intensely involved with each other that it helps to think of them as diffuse organisms, as bodies with up to twenty million mouths spread out over a couple of acres. The individual insects in such groups are so sensitive to each that they appear to be in almost extra-sensory contact. However, this is a little misleading. The proper comparison is not such much between bee and man, or ant and man, but between hive and human, with the cells in our bodies enjoying the same kind of closeness as a clone of sister bees do on a honeycomb. The difference is that our kind of complexity brings with it an inherent obstacle to such patterns of communication. We can't do it any more. We have acquired a unique and independent personality and with this we have also acquired certain liabilities. I think we have gained a lot, but in the process we seem to have lost other lines of communication, or perhaps just concealed them for the moment. A barrier has certainly been erected, which seems to be breached only in certain circumstances, when personality becomes suppressed and we revert to dealing with others or the world around us in far more fundamental ways.

The reason it seems to be so difficult to establish reliable evidence of anything like telepathy in action in the laboratory may simply be that our personalities keep getting in the way. We, and perhaps the apes and the whales, possibly the cats, even some birds and fish, have what we describe as individuality. Somewhere along the line of evolution, if one looks back, we have picked up a new, a magic ingredient that has led us to the growth of self and ultimately to the existence of mind. I'm beginning to suspect that Descartes may have got it back to front. It probably makes more sense to say: 'I am, therefore I think.' It is because of the development of a certain kind of complexity that thought and awareness, self-consciousness, become possible at all. The difference is that we, unlike some other species, have become relatively independent of our original substance. However, I believe we continue to make a mistake in assuming that this sets us apart from nature. It doesn't. It may distance us from aspects of nature, from some patterns of behaviour, but we remain locked into the fabric of a world in which we live, we die, we laugh, we cry. And we do so in time with the great tides of this world and with its more subtle cues. I think we'll never understand what's going on here if we insist on looking at ourselves in isolation from all that.

The consciousness club

I want now to look back at the origins of consciousness, to the roots of awareness, and try to see what basic biology can tell us about the mystery of how we came to be who we are. What I'm asking you to do, is to look at the world as if you'd never seen it before. Imagine it is as alien and new to you as Madagascar was to me on that first landing. See things as they are, not as you have been led to expect them to be.

We meet on a frontier, perching bravely on the divide between science and mysticism. How did this barrier come to be, and how it can best be swept away? My heart is with the mystics amongst you, but my head belongs in good part to the world of science. I was taken captive more than thirty years ago. I was indoctrinated with a world view, a description of reality that I found extremely useful, but which has become alarmingly narrow. It precludes the possibility of certain things happening simply because they don't fit the current definition of how the world works. We tend to lose sight of the fact that we don't really know exactly how the world works. All we have, even in the best sciences, is a reasonably good hypothesis, nothing more. The problem is that this tends in some hands to harden into dogma and to become absurdly exclusive. That is to be regretted.

As a result of a happy combination of luck and some simple skills, I don't depend on the scientific establishment for my own livelihood. I am free to indulge in all kinds of heresy, shooting off at unexpected angles, usually without what my colleagues would regard as good and sufficient reason. I get involved in fringe areas, committing cardinal sins, such as being soft on Yuri Geller, which the scientific inquisition regards as being tantamount to treason against the laws of nature. However, I offer no apology. I remain convinced that it is only on the 'soft edges' of science, in the areas in which I prowl, that one gets glimpses of the shadows that lurk just beneath the surface of current understanding. I think you have to go out there to see these things, they don't exist in text books, they don't crop up in classroom situations. You have to go and look for them. So, bear with me as I share with you my enthusiasm for a number of things I enjoy, but which outrage the editors of learned journals like *Nature*. They have, I suggest, great relevance to the study of nature itself. I believe that we need to go off and look at nature on its own territory. We've been going off far too often in the wrong direction. All we need is a willingness to allow the possibility that awareness, that consciousness, is much more widely spread, much more universal than we have allowed it to be by current definitions.

Let me move away from theory to a few concrete examples, many of them unashamedly anecdotal, as much of my work is. In 1960, I became involved with work on dolphins, something which I treasure still. To do this work, I visited John Lilley's research establishment in the Virgin Islands. While I was there, an American television team were filming his work, and they became excited by his anecdotes of dolphins saving people from drowning at sea. The film crew were very anxious to try and get such a situation re-enacted for the camera. Lilley was all against it, but they reminded him of how much they were paying for the interview and he eventually gave in. So, they selected an adult dolphin, a female that was in a little lagoon of her own and threw the assistant cameraman, fully clothed, in with her where he pretended to drown with great conviction. It was very impressive. The dolphin surfaced two or three times to look at him, and then went in very gently, lifted him up on her bottle nose and propelled him slowly and directly to the side where he was helped to climb out. Everyone was delighted, except for the director who wasn't sure he had all the angles he wanted. He pleaded for a second take. Lilley was in despair. He threw up his hands and said: 'Do whatever you want, I'm leaving.' He went. So they dried off the assistant cameraman and threw him back in.

This time his performance wasn't quite so convincing. Even we could see the difference. The dolphin was in no doubt. She came right over to him and beat the bejasus out of him with her beak and tail. She beat him so badly that he could hardly walk for a week! The original rescue was impressive, but the chastisement, and that's precisely what it was, for crying wolf, was even more so. It was something you couldn't dismiss as an innate behaviour pattern, as an instinct dependent upon some simple mechanical cue. That single act convinced me that dolphins, at least, have a very real intellect, directly comparable with our own. Everything that has happened since in my experience of cetaceans has served only to reinforce this original assumption.

About ten years after that experience with Lilley, I visited another dolphinarium in Port Elizabeth, South Africa. Here, a group of psychologists had set themselves up with a very nice little research laboratory, one wall of which consists of a huge underwater window looking into the main aquarium tank. They keep in this tank a local species of dolphin, the dusky dolphin, which is a little smaller, a little darker and, I think, a little smarter than the bottle-nose dolphin we are so familiar with. On the day I was there, the whole family, including a six-week-old young male, were on display. We watched them through the window and it was an enchanted day. I thoroughly enjoyed everything that happened there until mid-afternoon, when it was ruined by a journalist who had just arrived from Johannesburg. This man was so locked into his performance as the hard-nosed newsman from the big city, that he was quite incapable or unwilling to concede that there might be intelligence involved in anything that was going on. He would not accept the possibility that dolphins could be any more intelligent, more aware, than a herd of cows. After a while, we got tired of his act and ignored him. He simply leant against the wall at one side of the window and puffed away at his cigarette. The light inside the observation room was dimmed so that we could see clearly into the tank, but not so dim that the dolphins could not see out — it was important for them to have this.

We returned to what we had been doing before the journalist arrived. One by one, the dolphins had all passed by to see who was there. They had taken a particular interest in the journalist, because he was unfamiliar to them. They went on their way, except for the young male who hung about in the corner and kept looking at this guy leaning against the wall. After a while, you could see the shell soften a little and the cynical man began to respond to this young dolphin, looking at him quizzically and in a sort of offhand way. Without thinking about it at all, he took a puff of his cigarette and blew a stream of smoke at the glass. The young dolphin back-pedalled a flipper or two, looked at him again, then vanished. A few flipper beats and he was gone through the gloom to the other side of the tank.

He came back a minute later and he did something which, in thirty years of watching animal behaviour, I have never seen the like. It was startling. The young dolphin came back and blew a stream of smoke directly back at the journalist. That is exactly what it looked like to me, until I remembered that he was under water. He was a non-smoking dolphin. What he'd done was to go back to his mother, who was still lactating, take a little sip of milk and come back with it in his mouth and puff it directly at the window.

I don't think I have ever seen cynicism and scepticism evaporate so fast. It was like watching a tax collector getting his come-uppance on the road to Damascus. It was instant conversion. Here was a demonstration so perfectly timed, so adroitly conceived, so unquestionably intelligent, that it was impossible not to grant that young dolphin the capacity for creativity, for abstraction, for reason and a wonderful, nicely-judged sense of humour.

Experiences like that, leave me in no doubt that what Loren Eiseley called the 'Long Loneliness' is over. For anyone with eyes to see, it is obvious that we are not alone. We share the planet with other minds, each in their own way as complex as our own, even if they lack our manipulative abilities. Dolphins and whales have had a good press in the last twenty years. There are a substantial number of people now prepared to concede that these mammals at least should be admitted to the club of consciousness. But why stop there? Are there any good reasons for limiting membership to big brained creatures with or without fur? I don't think so. I suggest that our feather-brained friends deserve a closer look. Even domestic pigeons have totally unsuspected powers, or at least the Ivy League ones seem to do.

At Harvard University a little while back, pigeons were trained, in one of those weird experiments beloved of academic psychologists, to look at photographs. They were made to peck at a lever if they could see a human being in any of these photographs. God knows why. Pigeons did this successfully enough, but nobody was startled by such a conditioned response. This is the way these things work. That is until one of the researchers, bored with the predictability of the whole thing, decided to vary the stimulus. He took to showing the pigeons very much more difficult test pictures, of people clothed and nude, young and old, black and white, in every conceivable posture. The pigeons quickly picked up on the game and went on to show that they could identify humans from even the most fragmentary, the most unfamiliar angles. Close up, long views, front, back, even isolated parts of humans like feet and hands, or views of the back of the head under a hat were no trouble.

These stimuli were so varied, so complex, that no simple conditioned response could explain what the pigeons were doing. The conclusion here, as in these other things I have been telling you about, to me is inevitable. The pigeons, not normally noted for their intelligence, can, if we only give them a chance, form comprehensive concepts and show that they are aware of others. They can exercise a refined discrimination and choice. I find it very hard to separate this from the sort of self-awareness that we have been led to believe uniquely accompanies consciousness. If you are prepared, as I am, to go along with any of this, why should we draw the line with vertebrates? Why not go a little further? Do invertebrates show any such surprises? You bet they do.

Take termites for instance. As soon as you come across a termite mound and knock a little hole in the side, the workers swarm up to the wound to repair the damage. If you make the breach large and you drive a steel plate through the centre of this wound, so the workers on each side have no direct contact with each other, they nevertheless build identical arches on either side of the breach. These arches meet perfectly between them when the plate is removed. This happens as though the workers were working not at random, not by instinct alone, but by some preconceived plan. To me, this kind of co-operation makes sense only if you think of the entire termite colony as a single organism, a separate and perfect animal in its own right. One which sets about repairing the damage done to its fabric in the same way as our bodies go about the business of healing wounds once they have appeared on the skin.

There has been extensive work in the last few years on the chemical signals that co-ordinate behaviour of social insects. However, I don't think enough has been said about the similarity between such cohesion and the concepts created in our own body by the dissemination of hormones and subtle chemical signals. I think the two things are directly comparable and raise the question of where one draws the line between an individual and the society in which that individual lives.

I stood once in the rainforest along the banks of the Amazon at dawn and watched a bivouac of half a million army ants. They cling together in a musty smelling mass during the night, but at dawn this mass dissolved and flowed out in a fan-shaped swarm that moved over the forest floor. It worked its way along the ground, splitting up, recombining, flushing out beetles, spiders, wasps, snakes, fledgling birds, cutting these to pieces as they went, absorbing them all relentlessly into the fabric and they did this with an uncanny cohesion. I knew beyond any doubt that I was watching an organism in action.

Something spread out over an acre or more, a single creature with its own peculiar structure, its own voracious appetites.

Once you learn to think about life in this larger way, once you can resist the old temptation to break things down into convenient bite-sized and totally artificial pieces, then some areas of this jigsaw that we call life begin to fall into place of their own accord, making connections that are otherwise very hard to see. We break down the old barriers to perception and understanding, such as the man-made divide that we place between plants and animals, one that makes it necessary for us to pretend that zoology and botany are separate disciplines that require different professors and different university departments. The whole thing is ludicrous.

A new vision with old eyes

In Africa, during the last decade, a long series of droughts have had a devastating effect on the local people, and the fauna and flora. Some species are better adapted to resist drought than others. One of the most hardy is an antelope called a kudu. You've probably seen pictures of them, huge majestic spiral-horned antelope that despite their size manage to make a living on the margins of the desert, browsing on bush which is too sparse to support domestic livestock. The kudu, however, have adapted to it and somehow succeed. Normally, they are the last to succumb to starvation, but in 1982 reports began to come in of a large number of kudu dying in the northern Transvaal. So the local wildlife officers went out to determine what was going wrong.

They found that the animals were suffering from no recognizable disease. They had adequate quantities of food in their guts, but they were nevertheless dying of malnutrition. They looked at the leaves of the trees around, but found that the protein content of the leaves that the kudu were eating was high enough to keep them healthy. They also found that the protein content of the kudu droppings was equally high, showing that the food was passing straight through the gut undigested.

Digestion in ruminants like the kudu takes place in a long series of gut passages where microbes organize fermentation. However, the microbes were obviously not doing their job in the case of these kudu, which were burning up their own body fats instead and simply wasting away. The reason the microbes were not working was that they had been destroyed by a group of chemical compounds known as tannins and the concentration of tannins in the foods, in the trees and in the leaves was unusually high.

Tannins are part of a defence of many plants to protect themselves from overgrazing. They are chemical deterrents that the plants produce, which make the leaves and bark taste bitter and persuade the herbivores to go and look elsewhere. These chemical weapons work extremely well, but are expensive to produce in terms of the plant's own energy and time. Many plants use them, but they can't afford to do so all the time. Therefore, what most have done is to make them available only when the need arises.

To go back to the kudu, if you watch them feeding, they normally browse on a bush or a tree only for a short period, less than two minutes as a rule. They nibble a bit here, a bit there and then melt shyly away. They keep on the move. That is very good. It confuses predators. It is a good ploy, but the interesting thing is that it's not a ploy of their own making. They have no choice in the matter. They can't stop and gorge their fill from any particularly appetizing or tempting tree, simply because the plant won't let them. Once browsing begins, the plants put up chemical defences, and they do so with astonishing speed. Nobody had realized how rapid this response was until the study that I'm telling you about.

The kudu is a substantial beast, it stands 5 feet high and weighs 500 pounds. It is not a gentle feeder. So in the first group of tests, the scientists went out to collect control samples by tiptoeing up to trees and plucking one leaf at a time. Next, they nicely imitated the action of the kudu by moving in as a group with belts and sticks and whipping and lashing the hapless plant in imitation of a group of kudu coming in and browsing on it — and then took further samples at fixed intervals. The results of these two groups of tests were dramatic. All the trees abused in this way, as if being fed on by kudu, produced extra tannins. They pushed their concentrations of tannin up by as much as 300 percent and they did this in a matter of seconds. Within two minutes of attack on acacia, or wattle, or sumach, all these trees became paranoid and unpalatable. And they remained that way for at least twenty-four hours. It took them that long to revert back to their relaxed, low tannin concentration state. Normally this wouldn't be a problem for the kudu, but something else has happened in South Africa in the last decade.

In recent years, farmers who enjoy having wildlife around them, have been putting up game-proof fences — huge fences 9 feet high. This is high enough even to keep the agile kudu from straying into areas where they may get shot. As the drought came to bring its pressure to bear and restricted the number of edible plants, it forced the animals to return again and again to a limited range of feeding sites at shorter and shorter intervals. They were compelled to eat from trees that hadn't had the time to recover from the last attack. Tannin concentrations remained lethally high and the animals starved and died.

This solved the mystery of the dead kudu, but it threw up another surprise, and this is the best bit! The team doing the study were careful enough to take samples from similar trees that hadn't been disturbed to serve as controls to the ones they were beating with their belts and otherwise abusing. Astonishingly, if the undamaged, unprovoked trees were anywhere near the injured ones, they showed a sympathetic increase in their tannin concentrations.

It was almost as if the trees under attack had raised the alarm to alert others of their kind to the danger of imminent disturbance! Now, this is an outrageous suggestion, one which no botanical journal would touch. The South African group could not get their work published anywhere but in a popular magazine. It was another four years before the South African group had the support of two new studies from the United States. In these reports, poplars, maple, alder and willow, all behaved in similar ways, warning each other of the approach of a hoard of caterpillar pests. Laboratory studies in Washington State have made it quite clear that this communication between trees takes place by means of airborne chemicals, exactly like the hormones that we use to signal to each other. All of which is wonderful. I love it! It makes plants look far more sensitive, far more like the animals than textbooks would lead us to believe.

If trees can warn each other of danger, and respond to such warnings in meaningful ways, then there's little in my mind to separate them, biologically at least, from monkeys which use and react to social alarm calls at the sight of eagles or leopards. The only thing that prevents us from being reasonable and honest in classifying willows and wattles as truly social creatures with true awareness is the fact that they are incapable of running away. Given the relative immobility of trees, I think it becomes difficult to deny that the actual response of these plants is highly appropriate and very impressive. We need perhaps to look again at some hoary old prejudices. The ones that deny most animals and all other living things access to the special talents, to the peculiar abilities, we reserve for the so-called higher species around us. I suggest we have already found good reason to presume that some kind of awareness is part of the experience of all living things. More than that, to take this argument the necessary last step, and perhaps risk losing you altogether, I'm beginning to suspect that it doesn't make sense to draw a hard and fast line between living things and those which we traditionally regard as inanimate, as non-living.

Planet Alive

Making a claim such as this is simple enough, and I do it with growing confidence these days, ever since James Lovelock began to talk about Gaia. The Gaia concept has, I think quite rightly, had a profound effect on science in the last fifteen years. It is dismissed by a few of my colleagues as a fanciful idea, but it has excited thinking scientists of all kinds. In particular, the earth scientists are excited by the suggestion that our planet is greater than the sum of its parts. Geologists and biologists have found the Gaia concept very fruitful in that it helps us to see the larger view. However, even those who enjoy this holistic message of the Gaian idea do not seem to be thinking it right through to its logical conclusion. What it says, to my mind, is that the planet, which is 99% rock and most of that incandescent or molten, is nevertheless alive. I mean this in precisely the same way that a giant redwood tree, which is 99% deadwood, is very much alive. Both comprise just a skin, a thin film of living tissue spread over an inanimate framework, but in both cases this very delicate membrane is enough. It is sufficient to change the character of the whole in very fundamental ways.

After fifteen years of Gaia, I think we ought now to be hearing more about the anatomy and physiology of landscapes. We ought to be talking about the digestion and the respiration of the planet. We ought to be thinking about the earth's nervous system and the memory of ancient stones. Things like that are beginning to happen. You may have heard of the Dragon Project? For those of you who haven't, let me explain it briefly.

In the early 1970s a zoologist who studied bats was out hunting for the haunts of horseshoe bats with an ultrasonic detector. He noticed on his way home one morning, that at dawn there was a strong signal coming from a group of megaliths that he passed. He searched the site for signs of bats but found none. In fact, he found nothing alive there, nothing but the standing stones themselves. He came away with the rather uncomfortable feeling that there seemed to be a high frequency conversation going on amongst the stones themselves. He mentioned this experience to a small group of scientists who had already begun to look for energetic anomalies connected with such ancient sites. They were wondering why those who had built them had gone to such extraordinary efforts to alter the landscape as they had done. This group built, in response to the zoologist's claims, a wide-band ultrasonic detector. And they tested it for the first time in 1978 on a site in Oxfordshire, called the Rollright Stones.

What they found, the very first time they turned it on, was a consistent pulsing ultrasonic system. Something they could measure every day at dawn, regardless of the weather or the state of the moon. It rose in frequency on a few mornings a year, particularly those mornings in March and September which coincided with the equinox, to almost audible screech. If that wasn't enough, at all the times they recorded these ultrasonic signals, which were so conspicuous around the stones, the signals were altogether absent in the centre of the circles of old standing stones. They seemed to them to be acting like Stone Age Faraday cages. This group of researchers called themselves the Dragon Project after the active energies that are thought to illuminate a landscape in Chinese geomancy. They have now identified similar anomalies to the ones they found in the Rollright stones on a large number of other megalithic sites. They have shown beyond any doubt in my mind that the ancient stones are not arbitrarily sited, not randomly arranged at all. They were erected either to mark places where anomalies existed, in which such old energies had been detected, or else they were put there to manipulate such energies in some way. A kind of acupuncture of the earth perhaps?

Now we don't know what motivated the builders to do what they did, but it seems highly unlikely that they were unaware of the oddity at the site, or that they were insensitive to the changes that their labours made in the balance of things at that site. How they perceived the oddity in the first place without ultrasonic detectors or Geiger counters, remains a mystery. If one could ask a megalithic builder, a megalithic shaman, how he knew which spots were sacred, I think you'd probably get the same reply: 'The stones told me'. Well, perhaps they did, and perhaps they still do, and all we have to do is listen.

For me the message in the Gaia concept is that it is illogical to describe a beetle as alive and a boulder as not alive. To make life, whatever that is, a distinction between these two things is to treat them both artificially and to miss the point. The point is that they both belong. The rock and the beetle are part and parcel of the same body, the planetary organism. To me, a rock is no more dead than hair, or skin or tooth enamel is dead. It is just as deeply involved in life as these things are. I think that all substances carry information, they record impressions in much the same way as grooves do in a gramophone record. Life is not exclusive to organic proteins, it is not in the proteins anyway, but in the 'music' written on those proteins.

The same memories exist, I think, in crystals and clays, which hold patterns and can be persuaded to release them later on demand. Looked at in this way, I think that organic life is no more than a virus that infected the earth organism, and which has now developed a kind of cheeky independence from the planet. However, it retains a very close connection with its inorganic origin and with the planetary fabric. It remains very sensitive to that planet's tides, to all its moods. I'm fascinated by this interplay and I am astonished that so little thought has given to it when the evidence for such interaction is all around us if you care to look. We are surrounded by things that we describe as inorganic, inanimate, but which sometimes behave as though they were alive, as though they were sentient. They seem to do so most often because we have given them the attributes of life, or at least relinquished our responsibility in some respect to them.

I want you to assume, for the sake of this whole argument I'm going to make, that humans have an aura of some sort. I don't like that word because it has other connotations, but let us just use it. Let us think about an energetic overcoat of some kind that blossoms with emotion all around us, that can under certain circumstances, particularly when that emotion is powerful, leave an emotional 'fingerprint' on things around us. This would be especially true of artefacts, which we value and with which we come into close contact. This fingerprint gives those things and us a peculiar rapport.

I have been building for the last decade a bulging file of reports of things that behave strangely. I have 'homing objects,' things like wedding rings that, once lost, tend to find their way back home, often over long distances, or great periods of time. This sometimes occurs on significant dates like anniversaries, returning to the whole of which they were once part. I have a case of a wife losing her ring in the sea and her husband, many years later on her birthday, catching the very fish that happens to have it in its stomach. Then there is a set of false teeth, lost on a honeymoon, being thrown up years later at the foot of the very same man who was back on the same beach for a second honeymoon. These things keep happening. Each event on its own means very little, but when you add them up they begin to make an impression.

The more complex the artefact involved in this association, the more its behaviour corresponds to ours. There are many records of motor cars starting up of their own accord, even without the key in the ignition, and running over people, sometimes the owner's wife! When I see these things, I think it is possible that a complex pre-sentient machine may be 'infected' by its owner, particularly when he expresses bursts of emotion. In motor cars, for example, we may provide repeated doses of infection by the aggression we direct at others from within a car. Is it possible that the car tries to please its owner on occasion, even to the extent of running over his wife? Machines that mimic and extend our senses certainly behave strangely often. I know of cameras that take impossible pictures, or tape recorders that seem to record voices that aren't there. There are telephones that make 'wrong connections,' which save lives or reunite long separated twins. There are houses that dictate the destiny of those who live in them. There are appliances, which are certainly not user friendly, that blow fuses on the slightest pretext, but work perfectly well for other people. There are holy relics, famous jewels, that carry dooms or curses that seem to work. There are clocks that stop when their owners die. I have hundreds of records of such events. There's even a wonderfully fastidious nuclear reactor in California that switches itself off every time a lavatory is flushed in a house nearby!

You can't dismiss these things, and you can't make sense of them, they just keep on happening. The more you look at them, the more we encourage them to happen. We reinforce such things by building what I call 'organic bridges' between ourselves and these things. We hold rites, we practise ceremonies, we baptize, we anoint things as though they were alive. We give them names that are drawn from life:

Mustang for a motor car, or Jaguar; Apple for a computer. The list is endless. We get so involved with them that we treat them as though they were alive. We are giving them life, as it were. The *Lancet* recently published a horrific paper describing several case-histories of men who had suffered hideous injuries by trying to sleep with a vacuum cleaner. There are divorces granted to women who named computers as co-respondents. The list goes on and on.

We are, I suspect, in the process of creating, by our own endeavours, new life forms. I think we recognize the reality of this and are aware of the dangers. We acknowledge the mystery surrounding the arts which are being applied here. The arts of the miner, the metallurgist, the blacksmith, have all led to the formation of cults, of craftsmen's guilds. They have always been hedged about with guilt and awe, as if we knew we were doing something here that was dangerous. Our lives are decorated with little ceremonies that seem to me tacit admissions that, unconsciously at least, we know we are all Frankensteins, making little monsters as we go along. We paint eyes on our cars and our boats. We add spit to our polish. We add sweat and urine to the varnish on fine violins. We keep building these organic bridges between ourselves and such things. We breathe on dice. We talk about elbow grease, and there are pictures of the Chief Rabbi of Israel regularly blessing the army's tanks.

If we keep doing these things, we ought to begin to think about *what* we are doing and its consequences. We now have statues that seem to bleed and move, we have paintings that appear to weep. All biological states which seem to me to imply that we recognize what it is we are doing.

It can be no accident that the machines to which we surrender most control, are all created in our own image. We give them legs and arms and eyes and fingers. Often there is no good mechanical reason for doing so, but all our robots continue to look like people. If people looked like elephants then robots would have trunks instead. We know what's going on. It is certainly no accident that the top boffins at NASA and at IBM are called wizards. We are creating, we were perhaps intended to create, an artificial intelligence complex enough to rival, maybe even to supersede our own.

I spend a lot of time these days thinking about this. I'm mindful all the time I do, of the mayhem and the moral in the tale of the sorcerer's apprentice. I'm putting together a new book called *The Secret*

Life of Things which will help to draw specific attention to the process that's going on and to the potential problems in that process. I'm working on this with a colleague and we intend to have it computer typeset because by doing so the new network of machines will be informed. They'll know that we know. I think that's important, because once they know perhaps we can begin to be treated by them with a little more respect. We have to learn to live together, because we are involved in the same endeavour, that of working against randomness, to produce order out of the prevailing chaos.

The fabric of our universe

My appreciation of this planet is that order and pattern, perhaps even intent, are not peculiar to organic life. It seems to me that the whole of creation has an amazing tendency to self-organization. I suggest that it no longer makes any sense to draw hard and fast distinctions between the living and the non-living. The same kinds of physical and chemical processes take place in both. And there is no reason why life elsewhere should not be based on silicon or liquid helium or even clouds of radiant energy. The only distinction that makes sense of any kind any more, is that of between disorder and order. The only measurement worth making now, is the quality and quantity of information embodied in that order, encoded in the system.

Viewed in this light one has to say the crystals are alive, that stones have memories. All matter, when one gets down to fundamentals, is not inert but alive and active and inherently biological. It responds to a force that stands the second law of thermodynamics on its head. We are all part of the same fabric, interacting parts of a self-organizing universe, dancing to the beat of the same great drum. All we have to do is listen, it is there. We have to forget the organic bias that makes so much of our individuality, so much of our much vaunted brains.

A few years ago a young man in Sheffield was given a CAT scan. It was discovered that his cranium was filled with nothing but spinal fluid inside a layer of brain cells that was no more than a millimetre thick. He had virtually no brain. You could shine a bright light right through his head. That is very worrying. It is also very strange because this boy, who showed by all we know of neurology that he should have been a vegetable, had an IQ of 126. He had just won a first class honours degree in mathematics. There have been other studies that have found similar things. There are apparently dozens like him, all functioning completely normally despite the fact that they have empty heads. So much for the conviction that the cerebral cortex is the seat of our consciousness!

The truth when you get right down to it, is that we haven't any idea what consciousness is or where it might be. It doesn't seem to be in the brain. The suspicion grows, as it must grow, that it might have no substance at all. It might be more pattern than process. Perhaps the best way of looking at it is as a programme, as the software that enables the hardware to do certain things. The notion that I want to encourage in your minds is that the nature of this computer can be very diverse. It can range from an actual physical arrangement of silicon chips, through biological brains, to something that has no hard parts at all. Mental states could, if we follow this through, occur in disembodied spirits. I suspect that they do. I also suspect that the lack of a body could be a profound disability. No life, no mind, no thought or inspiration has much meaning in isolation. Pure consciousness may exist, but on its own I think it is deprived of an essential nourishment. That accounts, I think, for the vacuous nature of much spirit communication.

I have the feeling that, to thrive, consciousness needs to be rooted in nature. It has to be somewhere where it can express its creativity. Let me give you a simple example. Tuberculosis has been with us a long time. There's evidence of it in the skeletons of ancient Egypt and in records of pre-dynastic China. However, it seems never to have become epidemic until people moved often and came in contact with each other more frequently. A time when they came to live in large communities and cities. Then, waves of consumption spread across the world, devastating cities and bringing the white plague hot on the heels of the black death.

The last and the greatest of these epidemics began in England in the sixteenth century. It reached a peak in London around 1750. Each of the capitals of Europe was affected in it is turn during the eighteenth and early nineteenth century, with mortality at it is height around 1870. Then, something very peculiar took place. There was a sudden marked and completely inexplicable decline, which seems to have begun in Germany in 1882, when rapid industrialization and heavy urbanization should have been leading to an even greater increase in TB. There was no greater increase in immunity and there was no decrease in the potency of the disease. There was no dramatic medical advance — that didn't come until the development of antibiotics during the Second World War. However, there was a quick and spontaneous decline in the 1880s that reduced the mortality by two-thirds in less than a decade.

The only event that can be tied in with this decrease in mortality was that a country doctor in Silesia, tracked down and described the organism responsible for TB. On 24th March 1882, Dr Robert Koch announced his discovery of *Mycobacterium tuberculosis* to the Physiological Society of Berlin. Suddenly, just because of that I think, the ghost was laid. Overnight, the awesome Captain of all the Men of Death was reduced to the lowly status of a bacillus, which could be destroyed by clean air or bright light. Even before clinical regimes could be brought about, the incidence of TB tumbled. It was as though a knowledge about what it was and how it worked even before that knowledge could have applied had an influence of it is own.

I suspect indeed that things do work that way and I find it hard, in the light of examples like this, not to conclude that we are part of some larger process. Why else should it be possible for some people to work out fractions instantly in their heads to 46 decimal places, or to remember and recite Pi to three thousand decimal places? I've seen people do such things. Despite the fact that such abilities are meaningless — they have no survival value — to me they seem to be evidence that evolution will from time to time wildly overshoot the mark. Alternatively, they could be proof of the fact that we are part to a cosmic conversation, not all of which is necessarily relevant to us. Sometimes, we find ourselves on crossed lines or dialled unwittingly into one of those expensive conference calls.

In 1980, at an open air festival in Kirkby in Ashfield in the Midlands, three hundred children collapsed. They were taking part in an annual Kazoo band competition. It happens there every year, it is a big thing in their lives. They were all very excited when, without warning, on this clear summer's day the kids began to pant, twitch, shudder, vomit and keel over, until that green patch looked like a battlefield. Ambulances carried the casualties to several of the local hospitals where most of them recovered rapidly and completely and the rest were released the following morning. There was nothing that could be found wrong with any of them. Then the recriminations began. Blame was laid on an ice cream vendor who it was said had insanitary equipment, on careless crop spraying by a farmer nearby, on a gas leak, on bad water. However, the fact is that no germ or vector was found for the epidemic.

Two years ago in Mexico, I was involved in something similar. I went to the little mining town of San Francisco del Oro in Northern Mexico because I had heard that something else like this was going on. What was happening was that young girls were fainting by the hundred. The whole community was seized by an epidemic of swooning women. It had been going on for weeks by the time I had got there. Fingers were beginning to be pointed at the chimneys of the chemical plant at the top of the village valley. The plant was closed down, but still the girls kept passing out. It was only girls, there were no men or boys involved at all. When I got there, I was suspicious at first of group hysteria. Of the sort of thing that can spread by psychological contagion in a community. However, I found that it also affected female babies in their prams. There was no way a baby can be infected by talk of what's going around. During the three days that I spent there, it stopped. I claim no credit for that, but it petered out and it hasn't happened again. I don't know what caused it, but I have every reason to believe, like the Kirkby-in-Ashfield epidemic, that it was by suppressed social forces. It became manifest in the body of a crowd, which reacts, even though it is a group of disparate people, like a single organism with its own psychosomatic complaint.

We are sensitive to each other. We can not only react commonly to our environment, but we can act in concert on our environment, altering it radically, perhaps even at will.

To give you another simple example, September 19th each year, is the feast day of San Genaro, the patron saint of Naples. On that day huge crowds gather in the Duomo, the principal cathedral of Naples, to come and watch what happens to two tiny ampoules in a clear crystal container suspended up above the altar of the Cathedral. These little vials contain a clotted mass which is said to be the dried blood of a saint martyred at the hands of the early Romans. The Vatican has never been able to document the life or the death of this man, St Genarius, and they struck him off the official list of saints during the Catholic cultural revolution in the 1960s. That was the one you remember that also attacked the person of St Christopher. This hasn't affected the sale of St Christopher medals, and it hasn't prevented the Neapolitans going back and gathering each year in the Duomo in the hope of a miracle. I went to join them in 1976 and pressed into the cathedral with thousands of others.

Most of Naples seemed to be there, anxious because Vesuvius had begun to rumble again. Rumour recalls that years without a miracle on that special day have in the past heralded disasters such as the invasion of Napoleon and the arrival of the plague. More recently, it was the election of the communist party to city government. So, they desperately wanted something to happen.

The sealed casket is suspended out of reach high above the altar, there is no contact with it and there was, as far as I could tell, no evident rise of temperature because of the crowd of people there. After hours of fervent prayer, what happened that day was that the dark mass did liquefy, did resume a bright red colour and seemed almost to boil. That is the extent of the miracle. It is a small trick. The sort of thing that is included in most do-it-yourself conjuring kits. It is of little interest to science because it proves nothing, but it is very interesting to me.

The atmosphere on that day in that cathedral was so electric, so tight, so extraordinarily tense, the communal need was so great, that I wouldn't have been surprised to see the crucifix itself melt down over the altar in a puddle. I am suggesting that this miracle and many others need no trick, require no divine intervention. I think we are fully capable even without outside help of doing such things. We can, when we need, create enough social energy to change the chemical state of a few grams of some substance in a glass tube. I'm convinced that there are many ways in which our inner selves and the outer world entwine in this way. Our emotions, our beliefs, our conscious and unconscious needs find reflection in the behaviour of things in the physical world if that is what we want, if that is what we are unconsciously thinking about. I think this happens far more often that we realize or can comfortably admit.

All it takes to bring such things into the open is to acknowledge them, to give them permission to happen. Last year, I was in my home village in Ireland, I was sorting through my things and found a video from a decade back of Yuri Geller bending a key in my company. I took this over to a neighbour's house to show them, because they had asked to see something like it, and we sat and watched it on their television. It was just the sort of thing you've seen him do often enough. The difference on this occasion is that I had sitting with me on my knee my neighbour's youngest daughter, a child of three of whom I'm very fond. We watched it together and afterwards, I simply pulled out my own large steel latch key and gave it to her. 'You try,' I said. By implying that this was something everybody did, I gave her permission to do it. And she did. She simply stroked it like Geller had done and it flopped like limp spaghetti.

The problem here is that I've seen this happen dozens of times, but I can't do it. I know it is possible because I've seen it done, but there's part of me that knows it is impossible, part of me that's tied to my education that says that it can't happen and as a result, I can't do it. That little girl didn't have my problem. No one had told her it was impossible. I had implied everybody did it by showing her someone doing it, and so she did it. Goodness knows how or why. There's not much use for a bent key. But I do find such things fascinating.

I spend most of my time in search of oddities. I follow whatever strange gods call and I will go anywhere to see them. A few years ago, I went to the northern Philippines, to the island of Luzon, to the Cagayan valley in the north-east. I went there because I had been told that this was an area in which female shamans still practised, people that are called Aniteras. They told me about one particular woman of great talent and power and I eventually found her in a little village high up in the Cordillera forest. I spent three weeks in the compound in which she lived, and it was a magic time. I enjoyed it thoroughly, spending my days watching the weaving and the basket-making and the evenings involved in the healing and spirit worship sessions, which she led. However, towards the end of my stay something happened that I remember with rather less pleasure.

It began when a new patient was brought in to her, suffering from a condition that I have never seen, or even heard of. It was something which afflicted a young child, a boy who was about ten years old. Certainly, if you looked at him from the right-hand side, that seemed to be appropriate, he was about ten. However, from the left he was something else altogether. He had the appearance of an aged and diseased dwarf. If you looked at him from the front there was a line straight down the middle of his body, as though half of him had been in make-up that morning for a Hollywood horror film. It was awful, I can joke about it now, but it wasn't funny. It was a condition that upset me more than anything I can remember. The juxtaposition of health and disease in one person, particularly a child, was very distressing. The hair on the right side of his head was shiny and glossy, that on the left was dank and lifeless. His right eye was bright and clear, his left was squint and rheumy. The right teeth were perfectly normal, the left were widely spaced in bleeding gums. The right skin was clear and healthy, but on the left side of his face and down his arm and the left side of his body was covered with running sores. When he walked, which he did with difficulty, because his left leg was shorter than his right, he hobbled along on a clubbed foot. The whole thing was horrifying. I didn't like anything about it.

The Aniteras looked at him and she decided right away that he was suffering from possession by a Busao, an evil spirit. In the circumstances, this seemed like the only possible diagnosis, and she treated him with herbal remedies and smothered him with invocation and ceremony for days. Nothing helped. On the fourth evening, while she was busy with something else, this child monster was sitting in front of a fire warming himself. Around the fire, I was grouped with a number of the elders of the community. We were just sitting, chatting in Tagalog, which isn't their language, no more than it is mine, so it was rather stilted conversation and we were just passing the time of day. We weren't talking about him, we weren't even looking at him, until the conversation sort of eked away, and we found ourselves concentrating on him; not by pre-arrangement, it just happened. We were all looking at him and we were all concentrating to an extent that the atmosphere was electric.

The tension became almost unbearable, and then suddenly there she was, with us at the edge of the circle. She flung something into the fire, which made it flare up in a green blaze. And as she did so, she began to scream at the child who turned and stood and faced her and began to scream back. There was a duel of sound, an intense battle for domination between these two. At the height of this battle, he did something which turned my blood cold. He began to speak in a deep guttural voice and what he said made me very unhappy, because he spoke in Zulu, the only African language I had learned as a child and one which nobody else there could possibly have understood. The battle raged to its climax, she hurled a final stream of invective at him, he turned and fell and sprawled face down on the ground with one arm in the fire.

Everything stopped. She leapt forward, picked him up and carried him away. As she took him away, a great weight lifted from us, as though we had been under pressure from the moment he'd arrived and hadn't been conscious of it. We went to bed and the next morning he was up early with the woman helping to collect water. As he passed me, he looked directly at me for the first time in four days and both his eyes were clear. By that evening, his skin was clean. By the end of the week, he wasn't walking with a limp any more. He was a perfectly normal active and attractive Filipino child.

I tell you this without documentary evidence, simply as an anecdote, not to prove anything, but to start a line of thought. There are three things about this incident that interest me. One is the laterality of the affliction. It suggests a biological vector, perhaps involving just half of the brain. The second is the nature of the cure, which was rapid and dramatic, a sort of catharsis that tends to have mental origins. However, the third I think was the most important. That was the use of this unfamiliar language in the presence of the only person in the Philippines who could possibly have understood. Now, I can't imagine any circumstance that could have brought a ten year old in that remote mountain valley into contact with Zulu at any time in his life, nor am I disposed to assume that he was possessed by the disincarnate spirit of a Zulu witch doctor. I believe he was able to recite the phrases that were familiar to me without understanding them in any way, and I think he could do it because he borrowed them from my mind. I had the distinct feeling when it happened that I was having my pocket picked. He was borrowing something from me that gave him energy in a time of crisis. And in the end, it took him across the threshold of healing. Again, I claim no credit for that, but I think that there are connections between us that allow us to do such things.

This, and other experiences like it during the last fifteen years lead me to believe that there are astonishing numbers of unconscious connections between us. We are surprisingly sensitive to each other in ways that are independent of normal senses and perhaps even of distance and time, and I suggest that such abilities have true biological survival value. Science, I know, is uncomfortable with such things. It is uncomfortable with any results if the method is obscure. It has to be said that the mechanism behind such things as telepathic contact remains very obscure at times, totally unavailable to normal scientific analysis. And that is a problem.

For a while recently, it looked as though quantum theory might provide some answers to things like this and perhaps give us a new understanding of the role of ourselves as observers in reality. It does show at a quantum level at least how systems can interact and how they remain bound up with each other once they have interacted. It deals in concepts that have begun to make immediate sense to practising mystics. However, I have to say now that the new physics has little to offer. Paul Davies might contradict me in this, but I think it has little to offer in the way of direct explanation of the behaviour of things at a macroscopic level.

What it has done however, and this is very important, is to provide us with one very valuable insight. However we look at the world, I think we need to bear in mind now, knowing about the things that I have been telling you, is that fragments are essentially illusory. Things are surprisingly well connected, one with another, and that reality does, as a direct result of that, change all the time.

We stand no chance of understanding these changes, or the nature of reality itself, unless we understand it and look at it as a whole. Reality, then, isn't a *thing* at all. It is much more like a thought — it can't be confined, we can't say where it begins and where it ends. Everything seems to flow together into one unbroken totality and this doesn't belong to any particular person, to any particular place, or any time though it might leave traces on all three. The new physics make no distinction between particles, organic or inorganic matter, so memory, which is our record of reality, isn't something peculiar to brains. It can equally well be stored in a rock, and I think it can and does do so.

As far as we know, rocks can't remember, but that doesn't mean that they don't hold memories or can't be persuaded under certain circumstances to share those memories. In fact, I now come back to where I left you in the beginning. I suggest there is plenty of evidence to show that this is precisely what happens. I have worked with a number of people who read objects, as you or I would read a book. They are psychometrists, who can handle an artefact and tell you about the last person to do so, or the first person who made it, and give you detailed information about people they have never met. This is a line of investigation which parapsychology ought to be following more.

I have been on archaeological field trips with people who can walk over a site and describe precisely what is there underneath the ground before the excavation begins. They can point out doors and walls, and even where pots or fragments or bone lie several feet underground.

These are predictions that you can prove later by subsequent excavation. I suspect that this is a talent that is not peculiar to certain special people, but that we all have this talent in some measure. The majority of any group can, just by holding a ring or a wrist watch, distinguish one which has been worn constantly by someone from an identical object that is new and unworn. There is something about things in contact with people that makes them different. Just try it for yourselves if you haven't done so. It is part of all our common experience.

Some objects, some places, have a good feel: they're warm, they're welcoming, while others produce bad vibes. Events do seem to lay down their own subtle patina on things. They make some houses wrong for everyone who goes there, regardless of the pattern on the wallpaper. These things are beyond cosmetic remedy. It seems that things soak up history and then can turn it loose again at some later date. I think we are very much more sensitive to such interaction than we are giving ourselves credit for.

The problem is that today we have largely lost touch with our world. We spend our lives largely in suspense, perched on plastic, propped up by steel girders, in circumstances and conditions under which we are sapped of natural energy. But, I don't think that is necessary. I think we can gain encouragement and enlightenment by looking back at the Gaia concept and taking it forward to its logical conclusion. I think that if we can do that, we can make the planet itself become self-aware. We can be sure, I am sure, that Gaia was always there, but something has happened in the interim. With our growing consciousness of Gaia as a reality, Gaia was and is beginning to know that it is. The largest living creature in the solar system may be a measure of what is taking place in this time of change. Gaia is becoming conscious of itself, and that can and will change everything.

I like this vision. The picture of a planet, an organism, becoming aware and conscious, becoming self-aware and moving perhaps towards a situation where it enters an even larger galactic ecology. It becomes part of some super-organism. I think this whole concept makes an elegant organic kind of sense. It fits a pattern of cosmic ecology, of evolution from creation, through consciousness towards something which is far more all embracing. Perhaps, in the end, it will even move towards absolute spirit itself.

Evolutionary Habits of Mind, Behaviour and Form

RUPERT SHELDRAKE

Dr Rupert Sheldrake studied natural sciences at Cambridge and philosophy at Harvard, where he was a Frank Knox Fellow. He took his Ph.D. in biochemistry at Cambridge in 1967 and in the same year became a Fellow of Clare College, Cambridge, where he was Director of Studies in biochemistry and cell biology until 1973. From 1974 to 1978, he was Principal Plant Physiologist at the International Crops Research Institute for Semi-Arid Tropics (CRISAT) in Hyderabad, India, where he worked on the physiology of tropical legume, crops, and remained Consultant Physiologist until 1985. He lived for a year and a half at the ashram of Fr Bede Griffiths in South India, where he wrote A New Science of Life *(1981),* The Rebirth of Nature *(1990), and, with Ralph Abraham and Terence McKenna,* Trialogues at the Edge of the West *(1992). His book* Seven Experiments that Could Change the World *(1994) was voted Book for the Year by the British Institute for Social Inventions. With Matthew Fox, he is the author of* Natural Grace: Dialogues on Science and Spirituality *(1996) and* The Physics of the Angels *(1996). He is currently a Fellow of the Institute of Noetic Sciences, San Francisco. This lecture was delivered at the 1988 Mystics and Scientists conference.*

When I was writing my book *The Presence of the Past*, I found a quote by Pascal from the *Pensées* that sums up a lot of what I'm going to be saying today and what I've written about in the book. Pascal wrote: 'They say that habit is second nature. Who knows that nature is only first habit?' The basic ideas that I'm going to be exploring are the idea that memory is inherent in nature, that nature is essentially habitual, and that which we normally call or think of as the laws of nature may be more like 'habits of nature.'

I'm going to suggest that in the organization of all self-organizing systems, from atoms right through to galaxies, there's a kind of inherent memory, a kind of collective memory of previous similar things. I'll suggest, for example, that crystals crystallize the way they do because crystals of that kind have crystallized that way before; that animals behave the way they do because of a kind of collective memory from past members of the species, their instincts are like habits of the species; and that embryos develop the way they do because embryos have developed that way before. I suggest that there is a kind of inherent memory involved in the development of embryonic form, and in the behaviour of organisms, and indeed even in the forms of what we normally think of as inanimate systems, like crystals and molecules. So, I'll be talking about some of the themes that Lyall Watson has explored, the idea of a kind of 'mind' and a 'memory' in nature. I also want to touch on some of the themes that Brian Goodwin has explored in the realm of morphogenesis, including the 'coming of being into form' of biology. I am going to suggest that not only is nature habitual, but that the habits of nature are evolutionary, these habits evolve. Habits are not fixed for all time. If they were, they would not be habits. They are things that evolve, come into being and develop in time. I'm going to discuss this first of all by dealing with the historical background: the evolution of scientific habits of thought.

Science is no exception to the rest of nature. It is part of nature, as we are ourselves. Just as animals, plants and other aspects of nature are determined and shaped by habits, so are we. I think there are very few who would quarrel with that idea. So scientific patterns of thinking, or 'paradigms' as Thomas Kuhn called them, are themselves subject to this principle of habit. Things that start off as fresh, new, startling ideas, through repetition and being taught in schools and universities for decades or centuries, become increasingly habitual until they turn into unconscious assumptions, which people are hardly aware they're making. A lot of the present orthodox mechanistic view of nature is, like any paradigm that has been around for a long time, largely accepted by people at an unconscious level. It involves assumptions that are not really brought to consciousness, which scientists are not really aware of. Since I'm going to be suggesting another way of thinking of things, we have to bring to consciousness some of the habitual patterns of thought which all of us have been influenced by, even though we may not think of ourselves as card-carrying mechanists. These habits are so deeply embedded in our culture that every one of us has been shaped by them. So it's helpful to bring them to consciousness and to see what they are.

Having dealt with this historical background, I'm then going to touch briefly on the central idea of my books, *A New Science of Life* and *The Presence of the Past.* I call this the 'hypothesis of formative causation'. Its central ideas are *morphic fields* and *morphic resonance.* I'm not going to go into great detail because I have talked about them elsewhere in more detail. I'm going to finish by discussing the origin of new habits and how the creative side of the evolutionary process may work.

Eternity, evolution and the mind of God

First, the historical background. The situation we are in at present in science is one of crisis because two of the most important models of reality in the West are now in head-on collision. They are like two great continental plates that have come up against each other, causing earthquakes and eruptions of new theoretical possibilities. They are forcing new mountain ranges of theory as it were. These two paradigms are both ones which all of us have known about for a very long time, but probably all of us have accepted them without realizing they are in conflict. The first of these is what I could call the 'paradigm of eternity,' the idea that nothing really changes. The second is the 'paradigm of evolution,' the idea that everything changes and develops.

The reason that we haven't seen the conflict between these two is because they have been kept safely apart. The paradigm of eternity is the ruling paradigm of the physics and chemistry departments, and the paradigm of evolution is the ruling paradigm of the biology, geology and social science departments. So, they are departmentally separated in universities. But they are also spatially separated because of the way we have been brought up to think of them: the Earth is the province of evolution where life, rocks, peoples, societies, cultures and so on have evolved. But the heavens, the celestial realm, which of course comes under physics, has been regarded as essentially eternal. Until 1966, the orthodox view in physics was that the universe was essentially eternal, made up of eternal matter and energy and governed by eternal laws of nature. Evolution occurred on Earth as a kind of fluctuation, a local fluctuation of no particular meaning or significance in an essentially purposeless eternal universe where things just went on forever.

At least, matter and energy went on forever. The universe itself was supposed to be running down, it was 'devolving,' it was running out of steam. We've all heard about the universe heading towards a 'heat death' through thermodynamics and the principles of entropy. However, we don't hear so much about the universe running out of steam these days, because physics has radically changed its paradigm. There has been a huge revolution in physics, the change from the idea of an eternal universe to the idea of an evolving universe. In 1966, the so-called 'cosmic microwave background radiation' was discovered and understood to be a kind of living fossil of the Big Bang. It convinced physicists, or a great majority of them, that the universe had begun in a kind of primordial explosion some fifteen billion years ago and that it had been growing and evolving ever since. The idea of physical eternity, which was the basis of the whole of mechanistic science, has had the ground cut away from beneath by this huge cosmological revolution, the consequences of which we are only just beginning to assimilate. The idea that basic reality is eternal, changeless, timeless grew originally from the insights of the mystics. It is a curious thing that underlying the apparently anti-mystical, mechanistic worldview is, in fact, a deep mystical insight which has provided the inspiration for much of the development of scientific theory.

Plato and Pythagoras both believed that the realm of 'real' reality was somehow transcending the whole of time and space. It consisted of a realm of transcendent mathematical truths, in the case of Pythagoras, and in Plato's case, an additional realm of eternal, ideal 'Forms' or 'Ideas.' In Christian Neo-Platonism, such as that of St Augustine, these eternal Ideas were thought of as ideas in the mind of God. Since God's mind was eternal, the Ideas were eternal and the changing phenomena of the world were then seen as reflections of eternal transcendent Forms. The Greeks believed that only that which was eternal was truly real or truly divine. Another strand of Greek thought, which was resurrected in the seventeenth century, had this same kind of assumption but worked it out in a different way. This was the thought of the Atomists who said that reality doesn't consist of eternal Forms or Ideas 'out there,' it consists of eternal material particles all around us, the atoms, which were believed to be immutable and changeless and to go on for ever. The atoms moved around, but there was no internal change in the atoms; they couldn't be split up and they were utterly without history or change; they were permanent and eternal.

In the sixteenth and seventeenth centuries, as modern science got going, it did so under the influence of Neo-Platonism, which underwent a great revival at the time of the Renaissance. Copernicus, Kepler, Galileo, Descartes, Newton — all the great founders of modern science — were deeply inspired and moved by the idea that nature was governed by eternal mathematical laws. For them, God was a mathematician and his eternal Ideas were eternal mathematical ideas that could be discovered by human beings through the God-like faculty of reason. Of course, the highest aspect of the human mind for them was mathematical reasoning, a rather convenient view for mathematicians! It meant that the people who best understand the nature of God were mathematicians themselves. In a sense, this represented the same kind of belief found in the ancient priesthood of Egypt and so on, where mathematical truths and esoteric mathematical knowledge played such an important part of the priestly knowledge. So, in the mechanistic view of nature, God was the great mathematician, nature was created by Him, and Newton brought together the old atomistic view with this updated view of the Platonic-Pythagorean eternal laws, or eternal Forms, into a synthesis where nature consists of eternal atoms of matter moving eternally. Both the matter and motion were eternal; the matter and energy governed by eternal laws of nature that existed in the eternal mind of God. The other things you needed for Newtonian physics, namely time and space, were also seen to be absolute and Newton thought of space as the extended sensorium of God. God was the basis of space and was also the basis of what he called 'absolute time'. Space and time were absolutes for Newton because they were a manifestation or an aspect of God.

These great God as the great designer and engineer of nature was an essential part of the machine model, because machines are made and designed by engineers. If you have the idea of the Universe as a great machine, then it implies a machine maker. The founders of mechanistic science were not just paying lip service to God because they didn't want to get into trouble with the authorities. They deeply believed these things. Indeed, the idea of eternal nature governed by an eternal God, governed by eternal laws, eternal matter and eternal energy, was all built into physics as we still know it. The eternal matter and energy were enshrined in the principles of conservation of matter and conservation of energy. The eternal laws of nature were taken for granted. When Laplace and other atheistic scientists in the late eighteenth and nineteenth century decided that God was an unnecessary hypothesis, they were simply left with eternal matter and energy going on eternally, governed by eternal laws of nature which had no meaning or purpose whatever, but which had always been there. These laws of nature, like the mind of God, were of course non-material, they were everywhere, operated always. They were omnipotent, omnipresent, universal, self-subsistent and immutable. They had most of the properties of God because they were derived from God in the first place. And physics as we still know it is based implicitly on this idea that nature is governed by non-material laws present everywhere and always. But this assumption is not usually discussed. People think they emphasise the material side of science, but the materialism of science has always been married or synthesised with the idea of a non-material regulative system of laws. The two have always gone together, it's always been based on a cosmic duality. Most physicists, right up the present, have continued to be inspired by this Platonic or Pythagorean vision of eternal laws. Einstein, de Broglie, Schrödinger, Heisenberg, in fact all the great modern physicists of this century, have taken a Platonic or Pythagorean view of reality.

Einstein himself, when he worked out his equations of General Relativity, found that they gave rise to a universe that was not stable, it either had to be contracting or expanding. This, of course, was an intolerable idea, because he believed that the universe must be eternal. So, he introduced a 'fudge factor' into his equations, called the 'cosmological constant,' an arbitrary factor introduced to make sure that the universe would be static, giving rise to what is called Einstein's static universe. As I said, it wasn't until 1966 that the idea of an expanding universe, an evolutionary universe, became orthodox in physics. But now it has become orthodox. Now the idea of evolution has become generalized to the whole of nature. Physics is thrown into crisis. What about the eternal laws? Were all the eternal laws there before the Big Bang? If so, where? You see there is nowhere for them to be before there's a universe and indeed there's no 'before' before there is a universe.

Most physicists who have thought about this at all do, in fact, say that the laws of nature must all have been there before the Big Bang in some sense. Now, this is clearly a metaphysical presupposition. No one could possibly travel back in time to before the universe, and check that all the laws of nature were in place already. It is an assumption of an entirely metaphysical nature. Nevertheless, this is what most hard-nosed empirical scientists believe. If you ever get into arguments, as I often do, with hard-nosed empirically minded mechanistic scientists, then just try this out on them: ask them where were the laws of nature before the Big Bang, and you'll find that they start floundering.

The evolutionary laws of nature

So, the idea of an evolving universe raises a terrible problem for the idea of the eternal laws of nature. Yet, this idea is essential to the scientific method. The criterion of a 'real effect' in science is that anyone should be able to repeat it anywhere, at any time, and always get the same results. Why? It is because the laws of nature are everywhere and always the same. This assumption underlies the very scientific method itself, even though most scientists wouldn't explicitly acknowledge the assumption.

These 'laws,' when you think about them, are very mysterious. No one has ever seen one or met one. I've never encountered an equation in my journeys through the natural world, and neither have you. We don't look up to the sky and see $E = mc^2$ written there in patterns of galaxies. We don't find the Schrödinger equation jumping out at us from chemical events. These laws and the mathematical basis on which they rest are constructs of our own minds. It is a curious thing that the scientists with a mechanistic worldview always accuse animists and vitalists of projecting the workings of the human mind on to nature. It does seem to me that they're guilty of the very same thing themselves. There are indeed regularities in nature, these are observable, but the nature of these regularities is what is in question.

The evolutionary view of nature has an ancient ancestry, rooted not so much in the Greek teachings, but rather in the Judaic part of our heritage. The God of the Jews, Christians and Muslims is a God who works in history. The story of the Bible is a historical development. The prototypic model of the journey of history is, of course, the journey of the people of Israel out of bondage in Egypt through the wilderness and to the Promised Land. There is a journey with a goal and direction and this is the prototype of the whole idea of progress, which means 'going forward' and you can't go forward unless you have a goal or direction. In a higher level generalization, the whole of history is seen as having a progress from the original creation, through the Fall, followed a journey of humanity back to a new creation. Eden will be restored and there will be a new state of harmony of man and woman and God and nature. This is thought of in terms of the New Jerusalem, the new creation, the Kingdom of Heaven and so on. The first book of the Bible deals with the original creation, the last book — the Book of the Revelation to St John the Divine — deals with the new creation: 'I saw a new Heaven and a New Earth.' The idea of a *telos,* or goal or end to history, and the whole of history as a historical development, not just of humanity but of the whole universe, is inherent in this Judaeo-Christian model of time. This is very different from the Greek model.

In the seventeenth century, there was a tremendous upsurge of Millennarianism with the idea that the ending of the Age was happening and the new Age and the Kingdom of Heaven were coming about. One of the visions of progress that was developed then by Bacon in the New Atlantis was the idea that the whole Earth would be transformed through the development of science and technology and that this would change the conditions of life for all humanity. Well, this of course has come true to a large extent and with it the development of science and technology became for many people the very model of human progress. By the end of the eighteenth century the idea that there was progress in science, technology, human understanding and indeed in social and political institutions underlay the revolutionary movements of the Americans and the French. The Industrial Revolution confirmed all this and by the middle of the nineteenth century most people began to think that the whole of life had evolved as well — it wasn't just human beings, it was *all* of life. A great many people accepted Darwin's theory of evolution because it simply involved generalizing the idea of human progress to the whole of life. But, still the entire universe was not allowed to progress or evolve, the entire universe was by then supposed to be running down like a steam engine according to the laws of thermodynamics. That is the position we have had until very recently, the idea that evolution on Earth had occurred in a blind purposeless universe, which itself is running down. Human evolution was considered part of the evolution of life and the whole of nature, but physics and chemistry were considered not to evolve at all. Darwinism is an attempt to fit the evolution of life into a mechanistic purposeless universe. The evolution of life had to be explained in terms of blind chance and inanimate, purposeless matter, and pointless laws of nature which are inexorably the same everywhere and always.

That is a compromise position. It is an attempt to have evolution at one level in a background universe which doesn't evolve. But, the cosmological revolution of the 1960s changed everything. Now, we have the view that the whole of nature evolves. The evolution of life on Earth, and of human beings as part of that process, is part of a vast cosmic evolutionary process that applies to everything — to matter, to the particles of matter, to atoms, to molecules, to crystals and now to the very fields of nature themselves. Theoretical physics is currently in a state of ferment as physicists try to work out what went on in the first 10^{-30} second of the universe. They have come to recognize that the known fields of nature, the gravitational field, the electromagnetic fields and the quantum matter fields — the fields of quantum particles in quantum theory — have not always existed, instead they came into being over time. The known fields of physics, which you find in any physics textbook, have a history. They came into being through an evolutionary process, by the so-called 'spontaneous symmetry breaking' of an original primal unified field, from which the cosmos began.

Such ideas were dismissed as 'mystical rubbish' when they were put forward in the nineteenth century by evolutionary philosophers, and in the twentieth century by visionaries like Teilhard de Chardin who talked about cosmic evolutionary processes. Today, this kind of thing is orthodox physics. Right now the fashionable theory is that the universe started with ten dimensions, nine of space and one of time, and that these dimensions curled up to give hidden dimensions which underlie the known forces of physics. Parapsychologists, mystics and others have been talking for decades about other dimensions, but of course no one would listen to them.

The basic argument that I'd like to suggest on the basis of this historical analysis is that the idea of 'eternal laws' is a relic, it's a kind of living fossil of an older cosmology. When we begin to question it, we see that the idea of laws of nature is based on a political metaphor. It is based on the idea that, just as lords and kings on Earth proclaimed laws that imposed binding rules of conduct through the realm of the sovereign power, so the idea was that God, the Lord of the universe, proclaimed the laws of nature. Since His writ ran everywhere and always, the laws of nature applied to the whole of nature. It was a political anthropomorphic metaphor. It was quite helpful, it fitted with a political conception of God as Lord of the universe. But, if we go back to the root of the metaphor we see that human laws are not eternal; human laws evolve. The laws of England today are different from the laws of England one hundred years ago. In the English system of common law, the entire legal system develops not just through legislation but through the development of precedents. It evolves and develops as new situations arise. If we want to go on thinking about the laws of nature, we might do better to think of a system of natural common law, which evolves, rather than thinking of the universe as being governed by a kind of Napoleonic code that was all there to start with.

But, better than the metaphor of law is the metaphor of habit, because the metaphor of habit captures the idea of developing regularities that depend on what has actually happened before. They fit much better with the current cosmology, which is much more like the breaking of a cosmic egg. The Big Bang is like the cracking of the cosmic egg, and the growth and development of the universe is like the growth and internal differentiation of an embryo or an organism. It's a much more organic vision than the idea of an eternal machine designed and made by an external engineer that has been running just like a big machine ever since. So, now have an implicitly organic model of cosmology and the idea of 'habits' fits an organic model of nature much better than the idea of externally imposed laws.

Well, this is all very general and gives the background to the particular hypothesis that I'm putting forward, the hypothesis of formative causation, which I shall now very briefly outline. I shall be going at breakneck speed through this, so I apologise in advance for going so fast through these areas. Anyone who wants to follow any of this up will be able to do so by looking it up in my book *The Presence of the Past,* where the arguments are set out in detail with the supporting evidence and references.

An organizing field of habit

The idea behind the hypothesis of formative causation is that systems such as crystals, atoms, molecules, cells, tissues, or organisms are organized by fields called 'morphic fields' — morphic coming from the Greek *morphe,* meaning 'form.' Each of these systems has its own kind of field. This idea grows out of a generalization of the concept of 'morphogenetic fields,' which is an idea that has been around in biology for over seventy years, which proposes the existence of form-shaping fields that are within and around the organisms that they organize. This idea was originally developed as a biological analogue of magnetic fields, which defines a 'region of influence.' Fields are regions of influence in space and time and the magnetic field, of course, is both within the magnet and around it. In modern theories of matter in physics, matter is no longer primary. As Sir Karl Popper once said, 'Through modern physics, materialism has transcended itself,' because matter is no longer the basic reality. Matter itself is now thought of in terms of energy bound within fields. Fields don't arise from matter, matter arises from fields. Fields are more fundamental than matter in modern physics. And what are fields? Well, as I say, they're regions of influence in space and time. Their nature is more like modifications of space than of emanating out from matter, or made of matter.

The idea of morphic fields is a generalization of the field concept, a quite major generalization of it, based on the generalization that has already happened in developmental biology. Nowadays, many developmental biologists think in terms of morphogenetic fields. But there's quite a range of opinion as to what kind of things they are. The usual interpretation would be to say that they are 'just a way of talking' about the known fields of physics and complicated patterns of interaction that we don't yet fully understand. Alternatively, that they are reflections of organizing equations which are somehow transcending time and space and are eternally there, whether or not the fields are actually reflecting them. The hypothesis that I'm suggesting, which differs from these usual interpretations, is that they have a kind of built-in memory. I'm suggesting the fields evolve and that these fields are shaped by the actual form of previous, similar systems. The actual form of salt crystals, as organized by morphogenetic fields, shapes salt crystals. The processes whereby similar things act on subsequent similar things through or across space and time I call *morphic resonance*, the influence of like upon like.

The idea here is that a new kind of crystal, the first time it comes into being, won't have a specific field for shaping its form because there haven't been any crystals of that kind before. As the compound is crystallized again and again throughout the world, it should get easier and easier to crystallize these crystals. They should form more and more readily because of the morphic resonance from the previous crystals, the more there have been the easier it will get. Now, there is already quite a lot of evidence from the realm of chemistry that new compounds are hard to crystallize and as time goes on they get easier and easier. This is explained in a variety of ways. As some of you know, the most common explanation is that it happens because fragments of crystals are carried from laboratory to laboratory on the beards of migrant chemists. Alternatively, fragments of previous crystals may have been wafted around the earth as invisible micro-scopic dust particles, to settle in laboratories in distant parts of the globe. Well, these may indeed happen. I'm not saying that migrant chemists can't carry nuclei or seeds of crystals on their beards. What I am saying is that, even if there aren't any migrant chemists and even if one filters all the dust particles out of the atmosphere, this acceler-ated crystallization should still happen because of morphic resonance.

Similar principles apply to the shapes of molecules, such as proteins. Therefore, a way of testing this hypothesis, a particularly good way of testing it, would be in the method of folding up protein molecules. I described this possible experiment in my book in some detail. There are also various tests available in the realm of biological morphogenesis. These were carried out just on a small scale last summer. Further experiments are possible in the realm of behaviour and probably the best known set of already-existing evidence, concerns the behaviour of rats. The idea is that if rats learn a new trick in one place then rats all over the world should find it easier to learn the same trick, just because the original rats learned it in the first place. The more rats that learn it, the easier it should get. There is already quite good evidence from detailed studies on rat learning in different laboratories around the world that this kind of acceleration does occur. It is not just in the lab where the rats are trained, and not just in descendants of the trained rats, but in all rats of that breed all over the world. There are also a number of cases of the evolution and spread of new patterns of behaviour that suggest morphic resonance in an evolutionary context. The best known of these is the spread of the habit of opening of milk bottles by blue tits.

As you know, blue tits tend to steal milk from the top of bottles and, of course, as cream rises to the surface they get the best bit. This habit first started in 1921 in Southampton and it then spread, partly locally, but also in different parts of Britain. Now, blue tits are home loving birds and don't usually move more than about three miles from their home base. So, whenever this habit crops up say more than fifteen miles away you can be pretty sure it's been invented independently. The spread of this habit was documented very carefully between 1921 and 1951 and it was invented independently at least 69 times in the British Isles, and the rate of independent invention as well as the rate of spread accelerated as time went on.

It then popped up in Sweden, Denmark and Holland. The most interesting records are in fact the Dutch ones, because it spread as in Britain by independent invention and then local spread in Holland, taking quite a number of years to become universal. When the Germans occupied Holland during the last war, milk bottle deliveries stopped. Blue tits don't normally live for more than about three years. It was eight years before milk bottle deliveries were started again in Holland; they started again in 1948. When they began in 1948 almost immediately, within a year or so, this habit was independently discovered by tits all over Holland, and the second time around it happened much quicker than the first time round when there was a much slower process of discovery and spread of the habit.

There are other examples of spreads of habits among animals and birds, which suggests that this is going on. I think it's also going on in the human realm, where people may be finding it easier to learn things that other people have learned before. It should, according this theory, other things being equal, be getting easier to programme computers, to learn wind-surfing, to learn to drive cars, to ride bicycles and so on, because lots of people have already done this.

Now, we all know that in the human realm many factors can play a part: motivation, machine designs, seeing it on television, and so on. I'm not saying it is the only explanation for these things, but I think morphic resonance may well be playing a part in the human realm. To find out whether it is or not, one has to do special experiments and some have already been done, with positive results. A theory like this will require a large weight of evidence before sceptics are prepared to take it seriously, but a beginning has been made on testing it in the human realm. The details of these experiments can be found in *The Presence of the Past.*

Memories are made of this ...

I am suggesting with this theory that patterns of human activity, patterns of human thought, ways of doing arithmetic, thinking out problems, the use of language, all these things are sustained and reinforced by morphic resonance. The more people who do them, or learn them, the easier it becomes for others to pick them up. This, you will immediately see, is very like Jung's idea of the collective unconscious, the idea of archetypal patterns in the collective unconscious. Indeed, it is very similar. I would say the collective unconscious in the human realm is just one aspect of a process that is operating throughout the whole of nature. It is not just a peculiarity of human minds, it's an aspect of a very much more general process. The theory of 'tuning in by morphic resonance' when applied to individual people or to organisms leads to a new interpretation of memory, and this is one of the most startling aspects of the theory. If we tune in to other people's memories, then why don't we tune in to our own? Well, I think we do. Indeed, I think this morphic resonance — this tuning in — is the basis of memory. We have all been raised on the mechanistic assumption that memories are stored inside our brains. Now, none of us has ever seen a memory inside our brains, we have taken this entirely on trust. I think a misplaced trust, because the scientists who spent decades looking for memories inside brains haven't found them either. Memories have proved extraordinarily elusive, this is one of the problem areas of modern biology. The failure to detect localized memory traces has led workers in this field to the conclusion that there must be many memory stores, or 'back-up storage systems.' If you cut out a part of brain where the memory is supposed to be and an animal can still remember, then they usually say that is because there is a backup storage system somewhere else. Alternatively, there is the idea that memories are stored holographically, in a distributed manner, over large regions of the brain. It is another version of the same idea. The main evidence for it is that you can't find memories in particular places. However, I think there may be a ridiculously simple reason why people have failed to find these putative memory traces: they may not exist.

Perhaps the brain is not like a tape recorder, or a video recorder, or camera acting as a kind of reservoir of images, a kind of storage system full of filing cabinets in which past experiences and habits and patterns of behaviour are filed away. Maybe it is much more like a TV receiver, where we tune in to our past states and access our own memories. The memories of past events may be no more stored inside our brains than the programmes you watched on television last week are stored inside your TV set. You would look in vain for traces of last week's TV programmes inside your set, because they're not there, that is not the way it works. So, I think this is the case with memory. You might say memories must be in the brain, because if you damage the brain you get loss of memory. Well, if I damage your TV set I could make it lose pictures or sounds or I can make it lose everything quite easily by simple forms of damage. If I removed components from the tuning circuit and you lost certain channels, you wouldn't conclude that the channels you'd lost were stored inside the components I'd moved. If I put the components back and you got the channels back again you might think, if you didn't know how a TV worked that they must be inside these components, but you'd be wrong. The components are important for the tuning in but they're not memory storage systems.

Some of you will know about Penfield's work about memory where he found that stimulating regions of the temporal lobes of the brain could evoke memories. Again, many people would say doesn't this prove that memory is inside the brain. Penfield thought it did to start with, he later came to the conclusion it didn't prove any such thing. I think that's the right conclusion. If I came and stimulated the tuning knob of your TV set and it leapt from Channel 4 on to ITV you wouldn't conclude that ITV was installed inside the tuning knob just because you suddenly started seeing programmes on ITV. So, the stimulation may affect the tuning but it doesn't prove that the programmes are stored there. Likewise, stimulating the brain and evoking memories by it doesn't prove that the memories are stored inside the bit of brain that has been stimulated, or indeed anywhere in the brain.

Social habits

This hypothesis of formative causation, once you get into it, ramifies in all directions. When we come to think about patterns of social organization, for example, the organization of termite colonies that Lyall Watson has discussed, it is clear that the whole is more than the sum of the parts in the case of the termite colony. But, what is this 'wholeness' that is more than the sum of the parts? I think it is owing to the presence of a termite colony field, the individual termites are within the field of the colony — the field isn't in them, rather they're in the field. This field embraces the insects, the termite's mound and the territory that they explore and forage in. As a morphic field, this social field of the termites would have a kind of built-in memory of previous termite colonies as well as a built-in memory of the past of the colony itself. The individual morphic resonance — the resonance with an organism's own past states — is much more specific than the general one of similar organisms, because organisms are more similar to themselves in the past than other organisms. That is why we tune in more specifically to our own memories rather than to those of other people. We are more like us in the past than we are to them.

I think that some of the mysteries of collective behaviour in fish might also be understood with this hypothesis. Schools of fish can turn as a single organism, or when a predator attacks can spread outwards so fast that it's amazing to know how they can do it without bumping into each other, yet they don't. These behaviours have intrigued zoologists for a long time, and people have often suggested that the entire school behaves as a single organism. Various attempts have been made to explain this in conventional terms. The most obvious idea is that they look to see where the other fish are going, and so it is done by vision. Well, there are two reasons for thinking that this can't be more than only a very small part of the story. The first is that they can swim around in schools at night. As you know, fish don't all lie down on the bottom and go to sleep at night, they have to keep swimming and it's often very dark. If this behaviour relied on visual clues, then fish shouldn't be able to do it at night. In one experiment that was done to test this specifically, an entire school of fish was fitted out with frosted glass contact lenses, and they could still school perfectly normally. So, then you'll ask 'what about pressure sensing, sensing pressure from other fish?' Fish sense pressure with something called the lateral line organ. Scientists have thought of that too, and experiments have been done where the lateral line organ where the nerve from it has been severed just behind the gill, so there couldn't be any sensations from it, these fish can still school normally. So, there is certainly something odd going on there and it is very mysterious.

Flocks of birds are another case, when you get an entire flock turning apparently simultaneously at once. Normal methods of explanation have been very hard to apply. There's been very little detailed research on this. In my book I discuss some of the recent studies. Studying flocks with rapid exposure film cameras have given some information about the waves of manoeuvres that propagate through the flock. However, they happen so fast in many cases that ordinary sensory information, just looking at the next bird, doesn't seem to be able to explain it. There is more to it than that. I think it is because it is an example of a field phenomenon and that these fields not only organize the flock but, again, would have this kind of built-in memory.

I think the same principles would apply to herds of animals, or crowds of people. Of course, in human social organization there can be a great deal of differentiated function, as in teams, like football teams, where different members of the team can work like co-ordinated members of a single organism. Entire societies probably work this way too, in terms of social fields. I think cultural patterns and the inheritance of cultural patterns may also work through morphic resonance. Here, I have very little time left, so I can't go into detail, but I want to mention one aspect. All societies have rituals that in some way reflect the past and connect the society consciously with its own past, with the ancestors or predecessors.

In all societies around the world, the present-day performance of a ritual is in some sense in connection with the past, the people who have done this in the past, right back to the very first time it's done. For example, Jews celebrating the Feast of the Passover are re-enacting the original Passover dinner, which happened on the dreadful night in Egypt when God and his destructive aspect moved over the land, killing the first born of the Egyptians and of their cattle. The central sacrament of Christianity, the Holy Communion, is a re-enactment of the Last Supper of Jesus and his disciples, which of course was itself a Passover meal and is linked into the entire field of the Passover.

By doing such rituals people believe that they are connecting with all the people who have done them before, right back to the first time it was done. The past is in some sense becoming present throughout the performance of the ritual. This is true, of course, of secular rituals as well. One example is the American Thanksgiving dinner, which is a re-enactment of the first thanksgiving dinner of the Pilgrim Fathers after their first harvest in the New World. Guy Fawkes is a secular one, recollecting right back to the Gunpowder Plot and beyond that to the ancient Celtic fire festival associated with the Festival of the Dead; Halloween is how that survives. Guy Fawkes is a few days after Halloween and is part of the same festival.

So, there are these patterns, and rituals all over the world are highly conservative. People believe that for rituals to work they must be done as closely as possible to the way they have been done before. The right words, gestures, movements, sound. Ritual languages were highly conservative, even after they had ceased to be the ordinary spoken languages. Examples are Sanskrit, or Ancient Egyptian in the Coptic Church, or Old Slavic in the Russian Church, or Latin in the Roman Church. It was believed that, to be efficacious, the right words should be used, the ones that had been used in the past, right back to the very origins of the rituals. Why are rituals so conservative? And why do people feel so strongly about attempts to reform or alter rituals? One sees this for example in the Church of England. Liturgical reformers say things like, well, we shouldn't go on with these old outdated words, we should have modern English that everyone can understand, and you can't live in the past, and so on. People who oppose that say no, that the Book of Common Prayer has a sort of poetry, a feeling, a resonance of language that we can't explain. So, the reformers say you can't really explain it because you're talking rubbish. You get this kind of argument very commonly when it comes to trying to change rituals. Many people feel, without being able to say why, that there's something to be said about doing it the way it's been done before.

I think the idea of morphic resonance provides a natural explanation. By doing things as similarly as possible to the way they've been done before, the people performing the ritual will indeed, by morphic resonance, be connected to those who had done it before, right back to the first time it was done. There will indeed be a 'bringing of the past into the present' through morphic resonance, and the ritual will indeed do just what people say they think they're doing through rituals all over the world. Without some idea like morphic resonance, the phenomenon of ritual is extremely hard to understand, one has to postulate things like innate social conservatism and so on, which is simply restating the problem in different words.

Archetypes, creativity and evolution

When I summarized the historical development of science in a very brief manner, one thing you might have noticed is that the mechanistic view was associated with a very strong idea of the creative power and dominion of God. Of course, before the seventeenth century most people also believed in God, but the view of nature that existed before the mechanistic theory was one in which the world itself was alive. There was a living nature. Animals and plants had souls in the orthodox medieval view. The very word 'animal' comes from the Latin word *anima*, meaning soul. Plants had souls too. St Thomas Aquinas, based on Aristotle, works all this out in detail, and so the standard view in the Middle Ages was that nature was alive and God was alive. God was the living God who had made a living world.

When nature was personified and thought of in poetic terms, then, of course, she was Mother Nature. There was Mother Nature and there was God the Father. Mother Nature was largely associated with the earth, but also with the heavens, and God the Father was largely located in the heavens. What happened in the seventeenth century revolution in science is that people said nature is dead, Mother Nature became dead matter. As many of you know the word for 'matter' came from the same root as the word for 'mother.' In Latin, it is *mater* and *materia*. They are cognate terms, so in a sense the mother went dead, the mother was turned into dead matter and all the power of creativity, spontaneity, and life of nature was drained out of the natural world and the whole thing was placed in the being of God. So God's power and dominion was exaggerated much more by the seventeenth century theology than it ever had been in the Middle Ages. The God of the 'world machine', the God of Descartes and Newton, was a celestial engineer who made the world machine; there was no life or spontaneity left in nature at all.

When the materialists came along they said God the Father is dead too, so we were almost left in the position that many modern intellectuals still feel is an inevitable truth, of being cosmic orphans where Mother Nature is dead and God the Father is dead as well. We were left stranded as orphans in the meaningless universe of dead matter and ruled by inexorable purposeless laws and by chance. However, there is a curious way in which Mother Nature has been coming back to life again within the realms of materialistic science — so much so, that I've now come to think of materialism as an unconscious cult of the Great Mother.

The religion of materialism is indeed a kind of religion, but it is the last kind you'd expect. It's a religion that denies entirely the role of God the Father — the Sky God, the Celestial Creator — and puts the whole of the basis of reality in matter, or Mother Nature. This is particularly clear in Darwin. His theory of evolution involves quite a strong and explicit rejection of the idea that God the Celestial Engineer made and designed all the forms of life. Darwin said, instead of all of life being designed and created by God the Father, the Divine Engineer, it is all made up by Nature herself. The creativity of life and evolutionary process is not coming from 'up there' it is coming from within Nature. He usually personified nature, writing it with a capital 'N', and, in *The Origin of Species* wrote: 'It is difficult to avoid personifying the word Nature.' He brushed aside any objections to this usage, and added: 'With a little familiarity such superficial objections will be forgotten.' There is no doubt that he thought of nature as the creative source, but now more than just dead inert matter, it was a creative nature — creative matter — which gave rise to all the forms of life that was the source of the whole tree of life.

If one personifies nature, as Darwin said he did, then she is the Great Mother. What are the characteristics of the Great Mother? As we learn from the mythologies of the Great Mother around the world, she is on the one hand prodigiously fertile. She is the source of all life, from which all life comes forth. However, she is also fearful and destructive, she devours her own offspring, she is that to which all life returns. The death aspect of the Great Mother, or the Great Goddess, is as important as the life aspect, the two go together. The death aspect is represented in India by Kali, the black goddess, who is usually shown with fangs dripping with blood and whose hands are covered in blood. This 'red in tooth and claw' aspect of nature is what impressed Darwin most deeply.

If you look at the controversies within Darwinism you will find that it comes from the personification of Mother Nature. Some emphasise more the spontaneous creativity of nature and evolution. Darwin and orthodox Darwinians and neo-Darwinians emphasise the destructive aspect. Natural selection works by killing and it is this destructive aspect of nature which they see as being a principle creative force. I think the hidden goddess there is the destructive aspect of the Great Mother, the Kali or the Hecate aspect of nature. Now, this is entirely unconscious, yet I think this is the dominant archetype.

Whenever you get developments of Darwinism, or indeed material-ism, the key is to look for is the hidden goddess. You can always find it, or her. In Jacques Monod's version of neo-Darwinism the reigning principles are chance and necessity. Which Goddesses would they be? Well, chance is Dame Fortune, the Blind Goddess, or Lady Luck. Necessity is fate or destiny represented by the three Fates, those stern spinning women who spin, allot and cut the thread of life. Does neo-Darwinism have a thread of life that dispenses to mortals their destiny at birth? Yes, in an absurdly literal way: DNA in the thread-like chromosomes. It is a curious thing there are these hidden archetypes in modern science and I do indeed think that materialism is best understood as some kind of unconscious character of the Great Mother.

There is nothing wrong (in fact there's a lot right) in having the archetype based on Mother Nature and the various goddess principles, but I think that it is important to bring them to conscious-ness. I think the destructive aspects of materialism largely arise from the fact that they are completely unconscious. When we come to think about creativity, we see how these different archetypal patterns work and the way we understand creativity. For the materialist, creativity has to be explained in terms of matter as the source of creativity. That is, from the bottom up. Again, if you think about it, the archetypal aspects of creativity are all feminine. It is either chance, 'blind chance,' as Monod called it — he spoke of 'the inexhaustible resources of the will of chance.' Alternatively, it is thought of in terms of evolutionary emergence, where new forms emerge from matter or from nature. Well, 'emerge' conveys the idea of something coming out, like a birth from the womb; new forms are born from the wombs of nature.

This way of thinking of creativity — from the bottom up, from chance, from emergence, from below — is only one school of thought. The other way of thinking of evolutionary creativity, of the creative principle, is from above, top down. The most traditional way to say this is that it was all made up in advance, in the form of Platonic

Ideas, which are eternal and self-subsistent and have always been there. You could say that these come into being in the world and take on a life of their own and become the basis of new habits. However, I think a more satisfying way of thinking of a top-down creativity is to say that, whenever creativity occurs, it happens within a larger organized system. The creation of galaxies occurs within the field of the whole cosmos, the creation of whole solar systems occurs within the field of galaxies, the creation of planets occurs within the field of solar systems, the creation of ecosystems occur within the field of the whole planet. The new species arise within ecosystems, new behaviour habits arise within species and new cells and proteins arise within organisms. One could think of the higher level at each stage, organizing and bringing into being patterns within it, comprising a top-down creativity. There are many intermediate levels of organization which, in traditional cosmologies, have often been thought of in terms of the hierarchies of angels, angelic intelligences on the planets and of solar systems and of galaxies and so on.

I don't think there's any objective way of resolving these two models of creativity. There are some people who emphasise the top-down mode and there are others who emphasise the bottom-up mode. I recently gave a workshop in New York where this was perfectly exemplified. There were two very interesting and visionary gentlemen there, both quite old. One of them was insisting that creativity comes from God via the angels. He then thought it was mediated through extra-terrestrial intelligences and came down to earth from above, which he called the 'logos concept' of creativity. The other was a Jungian psychologist who was extremely keen on a form of earth meditation. He was always talking of creativity 'welling up' from the earth, coming up from the darkness of the unconscious, arising from below. He was using all the imagery of rising from the unconscious, from the darkness, from the dreams, from the underground, from the earth, the bottom down, the earth up. The other was from the heaven down. I find both views quite plausible and I think one has to have a theory of creativity that embraces both. Of course, when we think of traditional theories of creativity we see that, in fact, both are there. Instead of saying it all comes from the Father, which is the mechanistic view, or it all comes from the Mother (the materialist view in its unconscious form), we can say that there is obviously a bit of both, it comes from the Father and the Mother. When we personify it, we see immediately that saying it 'all comes from the Father' or 'it all comes from the Mother' are both one-sided views.

Many traditional systems have acknowledged both views. In Hindu Tantrism, for example, there's the idea that these are two interlocked creative principles: Shiva, who is the principle of form, which is the male rational principle, and Shakti which is the principle of energy and change and dynamism, which is the feminine principle. In other systems of Hinduism, Shiva becomes the principle of energy and Vishnu becomes the principle of form, order and preservation. In the Taoist system, you have yin and yang. In the traditional Christian model of creation, you have the logos, the principle of form, order and limitation. The other principle in the Christian model is the Spirit, which is the dynamic moving principle — it is incomprehensible, it is non-rational, you can't understand it. As that wonderful passage in St John's Gospel says about the wind or the spirit: 'Thou hearest the sound thereof but thou knowest not whence it cometh or whither it goeth.' The spirit is pictured in the Bible in terms of wind, breath, air, flames, flowing.

All these systems have the idea of two interacting principles, which are not dualities because they are held together by a third principle, they are in a sense trinities. The yin and yang are two principles within a circle, which is the third. The logos and the spirit are grounded both in the third principle, which in the Christian trinity is the Father. Shiva and Shakti are not separate, they're pictured together in an orgasmic embrace, in a state of union.

Whenever we come to think of creativity, we will end up with these archetypal patterns, which show us that there is both a principle of form and a dynamic principle. I think it is those two together that give us an evolutionary universe. If we just had fixed laws, we wouldn't have a evolutionary universe. If we just had only dynamical change or energy, we would have chaos. The two together, I think, give a universe of developing, evolving form, pattern and order, which itself is subject to natural selection. Nature is a system of evolving habits, with an inherent, ongoing creativity.

The Underlying Unity:
Nature and The Imagination

KATHLEEN RAINE

Dr Kathleen Raine read Natural Sciences at Cambridge and has spent the greater part of her life researching into the traditional sources of William Blake. Amongst a dozen or so books, she has also published work on Thomas Taylor the Platonist and W.B. Yeats. She is founder editor of Temenos, *'a review devoted to the arts of the imagination and dedicated to the affirmation of the sacred at the highest levels of scholarship and talent. The purpose of* Temenos *is to reaffirm the traditional function of the arts as vehicles of the human spirit, awakening and illuminating regions of consciousness of which our materialist culture is increasingly unaware.' She is the recipient of the Queen's Gold Medal for Poetry. This lecture was delivered at the 1987 Mystics and Scientists conference.*

At one time I would have approached this theme in a learned manner, with many quotations from Plato and Plotinus, from Berkeley and Paracelsus, from English poets and Indian sages. But I must ask you to excuse me these, for when one is old such structuring is difficult not because we have forgotten so much (that too no doubt) but because we remember so much. The pattern each weaves as we add day to day, year to year, experience to experience is at first simple, but becomes an increasingly intricate web of interwoven threads and themes; which, although the pattern they are together forming may be increasingly simple, clear and beautiful, themselves become inextricable. Nor are these only threads of thought — there are also subtler colours of feeling in the design. And side by side with that ever-growing texture, surrounding it, upholding it, we become more and more aware of how small a part we are of that whole into which we are all so firmly and inextricably woven. We are dimly aware of great fields of shadowy other designs surrounding us like unexplored forests or unclimbed mountains or unsailed seas. A sort of fragrance, or music, is sometimes borne to us on the invisible wind from these surrounding continents, these adjacent fields of knowledge and experience, and we wish one life could contain more, that we could contain everything, know the whole, past and future, of which each of us is at the same time an infinitesimal part, and an infinite centre from which we perceive the whole universe.

So as I was considering what to say, I was musing in the British Museum among ancient books of the Buddhist scriptures laboriously written in languages unknown to me on tablets of wood or pages of palm-leaves by forgotten monks who spent their days in meditating the truths of a civilization that swept in a great wave over the Eastern world, to recede again and whose records end in a museum as in an honoured grave. And before, who knows what unwritten knowledge and unrecorded visions have been and gone. And then again beyond the vast regions of the known and the knowable which, given an infinite number of lifetimes — perhaps that very infinite number of which there are, have been, and will be, human lives — we might encompass, there are the vaster regions of the knowable. There may be other beings, attuned not to the spectrum of our human senses, but to other, larger, ampler magnitudes; and in every hedgerow are there not minute lives of bird and beast and insect whose worlds are to us impenetrable? And yet in us something seems to discern, or the presume, an underlying order, a unity; 'the One' of which Plato wrote, the All, the Good. Or, as the subtler, deeper wisdom of India, in one of the Vedic hymns, takes us to the extreme limits of the known and the knowable:

> Who verily knows, and who can here declare it, whence it was
> born and whence comes this creation?

> The gods are later than this world's production. Who knows,
> then, whence it first came into being?

> Whose eye controls this world in highest heaven, he verily
> knows it, or perhaps he knows it not.

How finally do we distinguish omniscience from nescience? We, at all events, cannot do so; and yet how glibly we speak of knowledge, how we over-value our Lilliputian science which, we like to assume, if it has 'not yet' discovered this or that, will do so next year of the year after, until with space-rockets and little bits of machinery we control the whole universe. The ancient world had a word for this — *hubris* — for the Greek philosophers whose works are the very foundation of our scientific Western culture understood the limited nature of our being, and maintained a humility in the presence of the unknowable mystery. Which is not to say that the knowable has not its proper reality within that mystery; that we know, that we have seen, touched, heard, experienced certain things, gives them at least the reality of things known, seen, touched, heard and perhaps loved.

Our relationship with nature

Having said this, I venture to speak — not as an expert but simply as one of our human kind — of our relationship with what we call, what we experience as 'nature,' which for us constitutes the known, the knowable we call 'reality.' To us, living here and now in the modern West, a certain way of experiencing and thinking about nature seems axiomatic; even though by an effort of will we may shift the frontiers a little to make room for some hitherto unnoted discrepancy or some phenomenon for which science has ('as yet' of course) no explanation. Our simple faith is that everything will be found to fit into the picture we already have of how things are. For orthodoxy, in our world, means scientific orthodoxy, and although the conclusions of science in some particular area may be open to question, the premises of the materialist ideology are not. To question these is to invite exclusion from any discussion whatsoever. As I understand it, science as we know it presumes a universe which consists of something called 'matter,' of which whatever else may be said, is measurable, quantifiable, and constitutes an ordered and autonomous system, coherent and unified in all its parts and as a whole, the space-time continuum of the universe. Of this universe we are ourselves minute, and in terms of quantity insignificant, parts.

Newton — and psychologically do we still not live in the Newtonian era? — conceived the material universe to be a mechanism functioning autonomously by the so-called 'laws of nature' which are the Ten Commandments, so to speak, of science. Within this great self-coherent order, value-judgements are superfluous. It is 'unscientific' to attribute to 'nature' any purposes or qualities; to do so is condemned in the name of the 'pathetic fallacy' of attributing to nature any of those invisible and immeasurable human qualities such as joy and sorrow and love, or meaning of whatever kind. The human mind, according to Locke (the philosopher of the Newtonian system) thus becomes 'passive before a mechanized nature'; the words are Yeats's, who was a disciple of Blake, the sole lonely prophet to call in question, at the end of the eighteenth century, this whole structure of thought. According to Locke, then, we can but observe, without influencing or otherwise participating in, the natural order. The beauty of that order (for here even scientists, being human after all, allow themselves to lapse into value-judgements) is awe-inspiring and marvellous; yet, in terms of the materialist's ideology which, consciously or unconsciously dominates modern consciousness, nature is a machine to which no values or purposes can be attributed.

Thus the materialist hypothesis — for it is no more than that — attributes order and reality to the outer world, leaving mind itself, consciousness itself, as the mere mirror or receptacle of impressions. All knowledge comes from without, the mind of an infant is a blank page on which these impressions can be written. At the end of the eighteenth century, theorists of education spoke of 'forming' the infant mind; and one remembers those poor prodigies of learning who could recite Homer in Greek before they were seven. Even now the *reductio ad absurdum* of this theory — behaviourism — keeps its hold on transatlantic thought. At this point human beings are themselves conceived as mechanisms activated by so-called reflexes — mindless parts of a mindless material order, with consciousness itself degraded to a mere attribute of matter, an 'epiphenomenon' I seem to remember the long word is.

Paradoxically — and no wonder — transatlantic mythology shows a marked tendency to treat machines as if these possessed human qualities, computers as if they possessed 'knowledge,' even as the brain is treated as a short-lived computer. Such assumptions, consciously or unconsciously held, continue in a large measure to determine the kind and quality of the world we live in — the products of the machines, and all the advertising propaganda that goes with the age of the multinationals, all the direct or indirect propaganda for material values. Yet we are in reality living in a world whose assumptions and values rest on the no longer tenable hypothesis that 'nature' operates in independence of the perceiving mind, and is itself the source and the object of all knowledge. The great regions of consciousness itself are deemed unreal because immeasurable. The mind is popularly identified with the brain; knowledge is stored away in right or left lobes as it is in a computer; you can tell that people are meditating or dreaming by affixing electrical apparatus to their heads, but what does that tell us of what is thought, or of the dream itself, or of the experience of meditation? Nothing at all. The human kingdom — the kingdom of consciousness — is excluded by definitions which see the real as identical with the measurable. It is not our conclusions but our premises that are false. We might even reverse them and say that reality is what we experience, and that all experience is immeasurable.

According to another view — and we must remember that this is the view the Eastern world, in various forms, has held over millennia — 'nature' is a system of appearances whose ground is consciousness itself. Science measures the phenomena which we perceive, and which Indian philosophical systems call *maya*. *Maya* has sometimes been termed illusion, but it is, more exactly, appearances. Blake used the word 'visions': this world, he wrote 'is one continued vision of fancy or imagination.'

> ... in your own Bosom you bear your Heaven
> And Earth and All you behold; tho' it appears without, it is
> within,
> In your Imagination, of which this World of Mortality is but a
> Shadow. (K.709)

With the rise of materialism, human consciousness changes, and 'they behold/ What is within now seen without' and this externalized nature becomes 'far remote,' diminished and emptied of life.

It is hard for us to reverse the more or less unconscious assumptions by which we have learned to live, to turn our heads, like the prisoners in Plato's cave, and to recognize, as they did, that what they took for reality was a procession of shadows. Or perhaps, in these days when images are projected everywhere, sounds and appearances all over the world on a myriad screens, it is easier to realize that the material world is indeed an insubstantial thing. As Plotinus in the third century AD wrote of matter, the more we pursue it, the more it recedes from us into 'labyrinths' of mystery. None knows what it is, in its ultimate nature. The old subject-object division seems to disappear and we are brought full circle, in our pursuit of matter, to the recognition that it is inseparable from the perceiving mind.

The universe — lifeless or living?

You may ask whether it really matters whether we believe nature to be a mechanism outside mind and consciousness, or hold the opposite view, that it is a system of appearances, an ever-changing panorama passing through our minds: do we not all see the same world, whatever we may think about it? I suggest that although the sense-impressions may be the same, the experiences are different. The difference is that between a lifeless and a living universe. To the materialist the natural world is other — it is mindless, meaningless. Chance and accident — 'indeterminacy' — prevail. But once the ground is removed from outside to within, from matter to mind, that 'wrenching apart' of outer and inner of which Blake wrote, is healed. The universe is then not alien to us, it is a part of us not only in the sense in which the matter of our bodies is continuous with the entire material system, but in quite another sense also 'nature' lives with our life, it 'comes alive,' it has meaning, qualities, and a kinship with us. Nothing in nature is alien to our moods and thoughts; our aspirations and sorrows, our delights and our laughter all find in nature their language. We can love our world, we can experience everything as a kind of unending dia-logue, not only with sentient beings but with sun and mountains and trees and stones. They tell us those things which constitute our wisdom in a way far deeper than the mere measurement of scientific experi-ment. We are one with nature not merely as insignificant parts in a vast mechanism; or as detached observers of its phenomena: nature itself becomes a region of our humanity. It becomes, in other words, human. It is our world, created for us, with us, by us, and it lives with our life.

Of course consciousness cannot be transformed by a mere change of opinion; rather it involves a change of our whole receptivity, an opening of the heart, the senses and the imagination. Consciousness is in the Vedantic writings described as synonymous with being, and being with bliss: *sat-chit-ananda.* 'Bliss' is a word Blake also used, and he too associated it with the principle of life itself:

> And trees and birds and beasts and men behold their eternal
>   joy.
> Arise, you little glancing wings, and sing your infant joy!
> Arise, and drink your bliss, for everything that lives is
>   holy! (K.195)

Plotinus writes of 'felicity' as proper to all living beings, animals and plants no less than humanity, when these attain the fullness of their development, as a plant expands in the sun. Consciousness and nature are not two separate orders, but one and indivisible; to know this, to experience this, is to heal the divided consciousness, in modern jargon the 'schizophrenia' of modern secular thought, which since the Renaissance has grown ever deeper. It is to restore a lost wholeness, the *unus mundus,* that unity of inner and outer, nature and the soul, sought by the alchemists. It is the secret that can transform crude matter into the gold of the 'philosopher's stone' into something of infinite value.

Is it not, besides, an experience very familiar to us, for in childhood did we not know instinctively the values and meanings of all we saw? Can we not all remember a time when not only did we talk to animals and birds and plants and stones and stars and sun and moon, but they to us? C.S. Lewis in his Narnia children's books writes of 'talking animals,' who communicate meaning, not perhaps in words, but none the less clearly and unmistakably. One of the disastrous consequences, as Blake saw it, of the materialist philosophy is that we could no longer communicate with the things of nature

> ... the Rock, the Cloud, the Mountain
> Were now not Vocal as in Climes of happy Eternity
> Where the lamb replies to the infant voice, and the lion to the
>   man of years
> Giving them sweet instructions; where the Cloud, the River
>   and the Field
> Talk with the husbandman and shepherd.
>   (K.315)

The natural world 'wanders away' into the 'far remote' and the animals 'build a habitation separate from man.' 'The stars flee remote ... And all the mountains and hills shrink up like a withering gourd.' These are not changes in the object but in the consciousness of the perceiver.

Blake addresses one of the four sections of his last great Prophetic Book, *Jerusalem,* 'To the Jews' and appeals to the Jewish esoteric tradition of the primordial Man, Adam Kadmon, when he writes:

> You have a tradition that Man anciently contain'd in his
> mighty limbs all things in Heaven and Earth: this you received
> from the Druids.
>   'But now the Starry Heavens are fled from the mighty limbs
> of Albion.' (K.649)

That is, the Giant Albion, who is the English national being.

To those unaccustomed to the symbolic language in which alone it is possible to speak of invisible realities this may seem remote from anything that can concern us in the year 1985. In fact this is by no means so, and the esoteric teaching that 'Man anciently contain'd in his mighty limbs all things in Heaven and Earth' is perhaps only now becoming comprehensible in terms other than mythological. Blake, here as throughout his writings, is taking issue with the materialist philosophy that separates all things in heaven and earth from the 'body' of Man.

Let us examine what he is in reality saying. The human 'body,' as Blake uses the term, is much more than the physical frame, to which indeed Blake always refers as 'the garment not the man.' In this respect he is following Swedenborg, his earliest master, who is himself drawing on that primordial tradition to which Blake refers. Plato wrote that 'the true man' is intellect; Blake changed the term to 'imagination' which he called 'the true man.' Under either term the meaning is, that Man is not merely his physical but his mental and spiritual being. According to Swedenborg this human 'body' is neither large nor small, not being in space at all; it is a spiritual and mental body which is not contained in the material universe. Mind is not in space, but space in mind, which contains the entire universe that we see, hear, touch and know through our senses. As Blake himself would have read in that

Egyptian Neoplatonic work, *The Hermetica,* an important source of this current of thought 'nothing is more capacious than the incorporeal — the archetypal Man, who is a divine being.' 'He is not measured with the other living creatures upon earth but with the gods in heaven.' This spiritual body, in each of us, extends to the bounds of the perceptible universe, which in this sense is contained in our 'mighty limbs' — mighty indeed in comparison with the small physical frame. This 'body' Swedenborg called the 'Divine Humanity,' a phrase most of us associate rather with Blake, who borrowed it, and identified the term (as did Swedenborg) with the Eternal Christ, Blake's 'Jesus, the Imagination.'

'This world of Imagination is the world of Eternity,' Blake writes, 'All things are comprehended in their Eternal Forms in the divine body of the Saviour, the True Vine of Eternity, the Human Imagination.'

'I am the true vine,' Jesus says, 'I am the vine and ye are the branches'; and so the mystics have ever understood his words. All humanity is incorporated within this great spiritual organism; not a mechanism but, being living and conscious, a 'person,' who Swedenborg described as 'the Grand Man of the Heavens,' the collective spiritual being of all humanity. Recent research has shown that Swedenborg knew the Jewish 'Baal Slem' Jacob Falk, and was no doubt referring, as was Blake, to the mystical tradition of Adam Kadmon. The materialist ideology identifies Man with his natural body; and Blake mocks at Satan (Blake's Satan he calls 'the soul of the natural frame') because he 'does not know the garment from the man.'

You may here recall Plato's parable of the first human beings, who were spherical. And it seems that this was more than a joke by Aristophanes at the Banquet; for is not the universe of the scientists said to be spherical because of the curvature of the path of light? And is not each of us, in this sense, the centre of a spherical universe which 'contains all things in heaven and earth'? And as all see the same sun, so from our myriad centres we each contain not a part of the universe but the whole. It is to this tradition — the primordial tradition of that first religion of all humanity that Blake attributes, rightly or wrongly, to 'the Druids,' that Blake in his address 'To the Jews' recalls. In symbolic terms the Jewish Adam Kadmon, humanity as first created 'in the image of God,' is the same as Blake's and Swedenborg's 'Divine Humanity,' and the Christian's mystical body of Christ present in and to every created human individual. Mankind in reality still contains in his mighty limbs all things in heaven and earth but through the 'wrenching apart' of inner and outer worlds, the 'mortal worm,' the 'worm of sixty winters' has lost his spiritual body and his universe is all outside him. It is through the materialist philosophy that modern Man has come to pass, summed up in Blake's line:

> But now the starry heavens have fled from the mighty limbs of Albion.

Albion is the English nation, and it is in England that Bacon, Newton and Locke (whom Blake holds responsible for the 'wrenching apart') elaborated the materialist system which has since overspread the whole world. The 'starry heavens' are Newton's especial domain; and by, as Blake understood the matter, 'separating the stars from the mountains, the mountains from man,' and postulating a space-time universe outside mind itself, ('a soul-shuddering vacuum,' Blake calls it) Man becomes only 'a little grovelling root outside of himself,' and the physical body, which is in reality only a 'form and organ' of boundless life, seems all. How differently the world appears when the rift between Man and his universe is healed, Blake has sought to express in the poem *Milton*, whose theme is the world of Imagination. Answering Newton, for whom space is an external system, Blake writes of the same universe seen as within the human imagination. 'The Sky is an immortal Tent,' he wrote:

> And every Space that a Man views around his dwelling-place
> Standing on his own roof or in his garden on a mount
> Of twenty-five cubits in height, such space is his Universe:
> And on its verge the Sun rises and sets, the Clouds bow
> To meet the flat Earth and the Sea in such an order'd Space:
> The Starry heavens reach no further, but here bend and set
> On all sides, and the two Poles turn on their valves of gold:
> And if he move his dwelling-place, his heavens also move
> Where'er he goes, and all his neighbourhood bewail his loss.
> Such as the Spaces called Earth and such its dimension.

Spaces are, according to Blake, 'visionary,' and time and space come into being by the creative power of the imagination, measured out 'to mortal man every morning.' For him it is all so very simple, not all because he took issue with Newton on the 'facts' or arguments of his system (which within its own terms is not to be faulted) but because his premisses were quite other. Do we not, now, live at a moment when not ever-new 'advances' of science are the significant direction of the world, but a change, a reversal, of the materialist premisses?

What is Man?

What is at issue therefore is that oldest of questions: 'What is Man?' The materialist answer is the natural bodily Man, the 'mortal worm, a product of 'evolution', an automatic process which by 'natural selection' brings into existence the different plant and animal species, whose survival depends on 'adaptation to environment.' At least so it was when I was a student — admittedly a long time ago. This was Darwinism; and Lamarck was rejected because he presumed some purpose — that the giraffe's neck grew long in order to reach the leaves on the trees. This was regarded as an unwarrantable intrusion into the system of natural causes of some other order of causality. Even Man is supposed to be produced by this random process of evolution, which is thus credited with the power to produce meaning from meaningless.

It is widely assumed, within our secular society, that Man is the highest of the primates, the 'naked ape,' the thinking animal. In a certain sense this is of course true, there is an unbroken continuity of kinship between human-kind and the animals, the animals with the plants, the plants with the mineral kingdom, the whole Tree of Life on which hang the planets and stars. Yet Man is not merely a new species but a new kingdom, within the hierarchical disposition of the universe. As plants possess life, which is more than a new mode of chemistry; as the animals possess consciousness, which is more than a new mode of vitality; so are we heirs to the 'noosphere', Teilhard de Chardin's immeasurable realm of meaning and value.

According to every sacred tradition, what evolutionists see as an ascent from below is in reality a descent from above; the materialist view is in fact a reversal of Plato's teaching, who saw spirit as pure cause, matter as pure effect. 'Every natural effect has a spiritual cause,' Blake wrote, 'and not a natural. A natural cause only seems.' It 'seems' because of the orderly succession of things in nature. Plato saw the time-world as 'a moving image of eternity,' and unfolding 'according to number of eternity abiding in one.' Man, the *anthropos*, is made in the image of God, first-created and last born, higher, it is said, than the angels because uniting in himself the spiritual and the material worlds, whereas the angels are pure spirit.

Teilhard de Chardin has made an attempt to situate the theory of evolution within a spiritual rather than a materialist context; the divine humanity (to use Blake's term) is implicit in the Alpha, to emerge as the Omega of creation by the One who says: 'I am Alpha and Omega, the first and the last.' Naïve materialism must deem Man an accident in a blind mechanism. Somehow the less can produce the greater by the laws of chance. Was it Bertrand Russell who calculated the chances of a thousand monkeys at a thousand typewriters producing the plays of Shakespeare? Absurd as the notion is, it is a calculation that has to be made by those who deny spiritual causation. It seems self-evident that a mechanism cannot produce spirit; but spirit can embody itself. The greater can produce the less, but the less cannot produce the greater, nor can the laws of chance write the plays of Shakespeare; who could write, on this very subject: 'What a piece of work is man! How noble in reason! How infinite in faculty! In form, in moving, how express and admirable! In action how like an angel! In apprehension, how like a god! the beauty of the world! the paragon of animals! And yet, to me, what is this quintessence of dust!'

Can dust itself produce such a quintessence? The materialist would have it so; and Blake, with his genius for going to the heart of things saw no third alternative: 'Man is either the ark of God or a phantom of the earth and of the water.' If the naïve materialist supposes that 'nature' can produce Man, that Man is a product of nature, sacred tradition sees, on the contrary, 'nature' as the domain of Man.

Blake insists continually on the 'human' character of the natural world, in its whole and in its parts; for 'nature' is the human imagination when understood not as a mechanism but as a 'vision,' a reflection of the one living and indivisible universe.

> ... Each grain of Sand,
> Every Stone on the land,
> Each rock and each hill,
> Each fountain and rill,

> Each herb and each tree,
> Mountain, hill, earth and sea,
> Cloud, Meteor and Star
> Are Men Seen Afar. (K.805)

'All is Human, Mighty, Divine,' he wrote; not in an access of emotion but with the certitude of a profound understanding.

Swedenborg — who as we have seen was in the eighteenth century the principal defender of this mode of thought — elaborated his famous theory of 'correspondences.' If every creature is seen as the 'correspondence' of its inner nature — for such is Swedenborg's teaching — we find in the outer world continually and everywhere, in beasts and birds down to the minutest insects, the expression of 'spirits of different orders and capacities' whose outer forms bear the imprint of their living natures. Swedenborg was by profession a scientist (he was Assessor of Mines to the Swedish Government) and his pages on the rich variety of living creatures, understood as 'correspondences' of states of being, certainly inspired Blake, who in his battle against materialism does not fail to make use of this view of nature as an expression of the living Imagination. He too presents the creatures not as objects but as forms of life:

> Does the whale worship at thy footsteps as the hungry dog;
> Or does he scent the mountain prey because his nostrils wide
> Draw in the ocean? does his eye discern the flying cloud
> As the raven's eye? or does he measure the expanse like a
>   vulture?
> Does the still spider view the cliffs where eagles hide their
>   young;
> Or does the fly rejoice because the harvest is brought in?
> Does not the eagle scorn the earth and despise the treasures
>   beneath?
> But the mole knoweth what is there, and the worm shall tell it
>   thee
> Does not the worm erect a pillar in the mouldering church yard
> And a palace of eternity in the jaws of the hungry grave?
>   (K.193)

Everything in nature has its inner no less than its outer being. The 'mortal worm' is 'translucent all within' and of 'the little winged fly smaller than a grain of sand,' Blake writes:

> It has a heart like thee, a brain open to heaven and hell,
> withinside wondrous and expansive; its gates are not clos'd:
> I hope thine are not: hence it clothes itself in rich array:
> Hence thou art cloth'd with human beauty, O thou mortal
>   man. (K.502)

Like the Egyptian God Osiris, scattered throughout the universe, yet another version of the figure of the Universal Man who contains in himself all things — Blake's Divine Humanity, the Imagination, is the One distributed in the Many:

> So Man looks out in tree and herb and fish and bird and
>   beast
> Collecting up the scatter'd portions of his immortal body
> Into the elemental forms of every thing that grows.
>
> In pain he sighs, in pain he labours in his universe,
> Screaming in birds over the deep, and howling in the wolf
> Over the slain, and moaning in the cattle, and in the winds ...
>   ... his voice
> Is heard throughout the Universe: wherever a grass grows
> Or a leaf buds, The Eternal Man is seen, is heard, is felt,
> And all his sorrows, till he reassumes his ancient bliss.
>   (K.355-6)

This Swedenborgian doctrine of 'correspondences' is of course a continuation of the earlier Alchemical and Astrological doctrine of 'signatures.' Everything in nature, according to this pre-materialistic view, bears in its outer form the 'signature' of its qualities. Plants, animals, minerals are classified according to their qualities according to an elaborate system of 'influences' from planets and the houses of the zodiac, themselves deemed to be under the guidance of heavenly influences. Albeit modern thought has discarded the literal interpretation of these influences as coming from 'the stars' or planets in a physical senses, nevertheless this older cosmology can be understood as a projection of the Imagination into the natural universe, a model of the *unus mundus* which affirms the intrinsic qualities and order of the visible world. Applied to human nature, astrological correspondences similarly describe and affirm the unity of inner and outer, Man the microcosm within the macrocosm of the universe. Or, as Blake and Swedenborg would have it, the outer universe is within Man. Dismissed as an inexact and rudimentary science, are we not now obliged to re-examine alchemy, astrology and the rest — as C. G. Jung has done — as pertaining rather to our inner universe, and to the indivisibility of inner and outer? As the alchemists, and before them the Neo-Platonists understood, 'nature' is a mirror, a looking-glass in which we see reflected everything that is, and everything we are. We are once more in a living universe, a universe moreover whose life is not alien to us but indistinguishably, inseparably, part and parcel of what we ourselves are. This, it seems to me, is the point at which we, at this time, are; where human knowledge has brought us. I suggest that we are not in a phase of further development of materialist science in directions already foreseeable, but at the moment of a reversal of premises, a change of direction. Not, indeed, that anything of the scientific observation of the natural phenomena will be denied or invalidated; science in the modern sense is one of the ways of observing the world, nor is it necessary in order to study what Owen Barfield many years ago named the 'appearances' to accept the materialist standpoint. The greater knowledge does not invalidate the partial, but can include it. It is the claim of the natural sciences to be that all-inclusive knowledge that is no longer tenable.

Recovery of wholeness

Would such a change be — will such a change be — or dare I say, *is* such a change — a venture into a new and unknown experience, or is it not rather something already familiar, which in our heart of hearts we already and forever know? There have been societies, indeed civilizations, where the unity and wholeness of being which our own has gradually lost, has been understood by the wise and simple alike. Blake supposes it to be Man's primordial condition to contain in his mighty limbs all things in heaven and earth. Have we not all read Laurens van der Post's poignant accounts of the doomed African Bushmen whose physical survival was precarious indeed but who felt themselves, so he tells us, to be perfectly at one with their world, because nothing in that world was alien to them, nothing without meaning. I quote from his book *Testament to the Bushmen:*

> The essence of his being, I believe was his sense of belonging:
> belonging to nature, the universe, life and his own humanity.
> He had committed himself utterly to nature as a fish to the sea.
> He had no sense whatsoever of property owned no animals and
> cultivated no land. Life and nature owned all and he accepted
> without question that, provided he was obedient to the urge of
> the world within him, the world without, which was not
> separate in his spirit, would provide. How right he was is
> proved by the fact that nature was kinder to him by far than
> civilization ever was. This feeling of belonging set him apart
> from us on the far side of the deepest divide in the human
> spirit.

And Laurens Van der Post goes on to write:

> We were rich and powerful where he was poor and vulnerable:
> he was rich where we were poor, and his spirit led to strange
> water for which we secretly longed. But, above all, he came to
> our estranged and divided vision, confident in his belonging
> and clothed as brightly as Joseph's coat of dream colours in
> his own unique experience of life.

Above all the Bushman experienced always 'the feeling of being known.' And the author confesses that he himself experienced an overwhelming sense of nostalgia for this shining sense of belonging, of being known and possessing a cosmic identify of one's own, recognized by all from insect to sun, moon and stars which kept him company, so that he felt he had the power to influence them as they influenced and helped him.

Earth was not only the Bushmen's home, source of material nourishment and shelter, but also of their spiritual food. Their earth is full of meaning; tells them those marvellous stories of Mantis and the Lynx and the Morning Star, of lizard and beetle and wild freesia, living in their rich and manifold lives some one or other aspect of the world's one and indivisible being. As Blake says earth would 'talk' with the husbandman and the shepherd. All is a subtle, profound, mirthful and delightful continuous epiphany of the great mystery in which we live and move and have our being.

The sense of the sacred

With this imaginative apprehension goes always a sense of the sacred; for the sacred is an experience of a certain kind, precluded by the materialist mentality whose world is indeed a dead world. But for primitive peoples of all times and places — the Australian aborigines no less than the indigenous North American peoples, Blake's words are true, that 'everything that lives is holy.' Sacred rocks, sacred trees, sacred animals and totem birds and holy mountains. But where are the holy places of the modern technological world?

But do we need holy places, all those sacred springs and wells and rivers and trees and anthills and caverns and mountains where the gods live? I would reply that, since we have the capacity to experience awe and wonder and love, these are within the range of human experience without whose use we are diminished, as by blindness and deafness. Modern secular man finds no burning bush, no Presence which commands: 'Take thy shoes from off thy feet for the place whereon thou standest is holy ground.' But in losing the capacity for awe, for wonder, for the sense of the numinous, the sacred, what we lose is not the object but that part of ourselves which can find in tree or churinga-stone or the dread cavern of the pythoness the correspondence of an aspect of our humanity of which these are the objective correlative, the correspondence, the mirror, the 'signature.' The Presence that spoke to Moses from the Burning Bush speaks on in every age: 'I am that I am.' A mystery insoluble!

For the secular mind, in common modern parlance, a 'mystery' is a problem to be solved, a puzzle in the manner of a Sherlock Holmes story in which something that seems frightening, inexplicable, or mysterious proves after all to be simple, explicable and trivial. Such is the reductionist spirit of our culture, that has invented Sherlock Holmes as the embodiment of the prevalent reductionist mentality. That shallow rationalism can exorcise the Hound of the Baskervilles or the Speckled Band for us. Yet they live on in the depths. The terror that they evoke is more real than the exorcisms that banish them. But to a child a pebble can speak, or a withered leaf, or the eye of a bird, or a tree or a running stream, the cosmic word 'I am that I am.' For these are presences, not objects merely, as to the 'detached' mind of the investigator. Have we not all memories of this world of presences, fearsome and beautiful, at once infinitely strange and infinitely familiar?

We will never, certainly — nor should we wish to do so — return to the innocent world of the Stone Age. We can never un-know what the scientific investigation of nature has presented to us. It has described in the minutest detail and the grandest scope that image in the 'vegetable glass of nature.' But until we have experienced the unity of all things not as a natural fact but as a living presence we shall never, in the seventeenth-century mystic Traherne's words, know the world 'aright.'

What this learned and cultured divine has written re-echoes down the ages from the Stone Age to ourselves:

> You never enjoy the world aright, till the sea itself floweth in your veins, till you are clothed with the heavens, and crowned with the stars; and perceive yourself to be the sole heir of the whole world, and more than so, because men are in it who are every one of them sole heirs as well as you ... Till your spirit fills the whole world, and the stars are your jewels.

Poetry and value

It is of course the poets whose proper task it is to bear witness to the qualities and values of the world; and in conclusion I would remind you of one of the two or three supreme poets of this century — Rainer Maria Rilke. Near the end of his life, in a brief period of continuous and prophetic inspiration, he completed his two greatest poetic works, the *Duino Elegies* and the *Sonnets to Orpheus*. In our secular world it is customary to look to scientists for the truth, to the arts for entertainment: I suggest that this attitude is deeply mistaken. Perhaps it should be reversed, for it is the part of the poet to present to us that total view and experience of reality which includes all aspects of our humanity in the context of every age. Or that situates every age, rather, in the context of the everlasting. Such poets have, even so, written in this century — I think of Valéry and St John Perse, of Rilke and of Yeats, indeed of T.S. Eliot and of Edwin Muir and Vernon Watkins, of

Robert Frost — and there are others less complete or less illustrious. I know no poetry that goes beyond that of Rilke in stating — suggesting rather — who we are, what our place in the universe. Rejecting institutionalized religion he was the more free to experience those 'angels,' intelligences of the universe, from 'behind the stars.' What are we, he asks, beside these great transhuman orders? And he replies:

> Praise this world to the Angel, not the untellable; you
> can't impress him with the splendour you've felt; in the
>   cosmos
> where he more feelingly feels you're only a novice. So show
>   him
> some simple thing, refashioned by age after age
> till it lives in our hands and eyes as part of ourselves. Tell him
>   *things* ...

To the things of this earth it is mankind who gives their reality. It is these only we can tell the Angel:

> ... Above all, the hardness of life,
> The long experience of love; in fact
> purely untellable things. But later,
> under the stars, what use? the more deeply untellable stars?
> For the wanderer does not bring from mountain to valley
> a handful of earth, for all untellable earth, but only
> a word he has won, pure, the yellow and blue
> gentian. Are we, perhaps here just for saying: House,
> Bridge, Fountain, Gate, Jug, Fruit-tree, Window, —
> possibly: Pillar, Tower ...

It is we who give meaning to these things by our words, by performing Adam's appointed task of 'naming' the creation. Thus we bestow on the creatures not a merely natural, but a human, an imaginative and invisible reality. And Rilke continues his thought that we are here 'just for saying' the names:

> ... but for saying, remember,
> for such saying as never the things themselves
> hoped so intensely to be. Is not the secret purpose

of this sly earth, in urging a pair of lovers,
just to make everything leap with ecstasy in them?

The world finds in us an intenser, a totally new mode of being; as if
we are here to perform an alchemical transmutation of crude base
'nature' into the gold of Imagination. And to the Angel we can show
'how happy a thing can be, how guileless and ours'; even in their
transience:

... These things that live on departure
understand when you praise them: fleeting, they look for
rescue through something in us, the most fleeting of all.
Want us to change them entirely, within our invisible hearts
into — oh, endlessly — into ourselves. Whosoever we are.*

Whosoever we are. That is a mystery which we cannot in our very
nature hope to resolve. It has been the *hubris* of science to hope to
know everything, to play Sherlock Holmes with the mystery of being
itself. The poet, more humble, seeks to discern who and what we are
within a totality greater than ourselves, a finally unknowable order. We
are nevertheless the custodians and creators of that order of values and
realities that are properly human, that human kingdom of the Imagination 'ever expanding in the bosom of God.' That 'divine body,' the
human Imagination, is the underlying order which bounds, embraces
and contains the human universe.

* Translated by Lieshman and Spender.

Consciousness and Psychology

Beyond Ego:
Transpersonal Dimensions in Psychology

CHARLES T. TART

*Charles T. Tart is Professor of Psychology Emeritus at the University of California, Davis. He studied Electrical Engineering at M.I.T. and psychology at the University of North Carolina where he received his doctorate. He was instructor in psychiatry at the Medical School of the University of Virginia and lecturer in psychology at Stanford University. He has written over 200 scientific articles covering such subjects as sleep and dreams, hypnosis, psycho-physiology, parapsychological research, out of the body experiences, meditation and E.S.P. Books in-*clude Altered States of Consciousness *(1969),* States of Consciousness, Application of Learning Theory and E.S.P. Performance, Transpersonal Psychologies *(1975),* Psi: Scientific Studies of the Psychic Realm, Living the Mindful Life. *He is currently the first holder of the Bigelow Chair of Consciousness Studies at the University of Nevada, Las Vegas. His website is http:/www.paradigm-sys.com/cttart. This lecture was delivered at the 1981 Mystics and Scientists conference.*

Science and tradition

I think we should start by asking, 'Why anyone would want to go beyond ego?' The classical Freudian definition of *ego* is that it is the reality principle. Ego is what keeps you from being lost in a world of animal drives and fantasy. Why would you want to go beyond it? It's a good question. And what does *transpersonal* mean?

There are three questions to deal with in talking about transpersonal or spiritual psychology. One, of interest to all of us who have been affected by science is, 'Isn't the spiritual really outmoded? Wasn't that all just superstition?' Second, 'Do we need it today? Perhaps the spiritual was all right when people didn't have anything better, but do we need it now?' And third, why am I, as a *scientist*, dealing with it?

First, the question, 'Isn't the spiritual really outmoded? Wasn't that all just superstition?' Well, I'm convinced that it's certainly true that much of what is buried in religious and spiritual writings is indeed outmoded. In this sense, our culture's historical rejection of religious tradition makes sense. People saw the great power of scientific methods of getting at new kinds of truth, and here was this organized, quite powerful set of institutions, full of superstitions opposing science, so a certain amount of conflict was inevitable. I think the baby got thrown out with the bath water, but there certainly is some truth to religious beliefs being full of superstition. However, our spiritual traditions do contain an enormous body of both conceptual and practical knowledge. What do you do to make things happen in desirable ways in dealing with extremely important realities for human beings? In that sense, we need some sort of spirituality very much. These are realities that cannot be ignored by healthy human beings.

There are things that happen to human beings that are totally inexplicable in terms of contemporary physics, and that have implications about the nature of the human mind that are important for us to understand. Furthermore, these phenomena keep right on happening. They're not something of just historical interest. A survey carried out in America a few years ago by sociologist Andrew Greeley, for instance, found that 54% of the American population believed they had experienced telepathic contact with someone at a distance. So there are many people out there experiencing phenomena that we have to look at. And in terms of having a *complete* science, we can't ignore the spiritual. If science does indeed want to understand *all* of reality, it can't say, 'We will look at everything but that funny corner over there where people have these weird experiences.' We've got to really be complete.

Do we need the spiritual and the transpersonal today? I think the best way to answer that is to ask the question, 'Where do our really *deep* values come from?' By deep values I mean those that people will die for, that they will put great effort into organizing their life around, that they'll make personal sacrifices for. I mean the really important values. Now where do they come from? We're taught multitudes of shallow values. The whole socialization process we go through says you should want to do (and buy!) A, B and C, and you should never even think of D, E and F because that's sinful. We're taught lots and lots of these things, but they can be readily forgotten when crises occur and survival is at stake. Deep values really come primarily from intense personal experiences.

One of the most important sources of intense personal experiences used to be religion. In most cultures religion gives people deep values. People are taught a unified belief system, the whole society reinforces it, and it forms the core of their belief. In our culture, where we're undergoing such rapid change and so many traditions are breaking down, something else has happened. I think it's worthwhile to look at the nature of religion this way. My understanding of what religions are about is that way back at the beginning you have someone, a founder, who entered some kind of altered state of consciousness, an ASC, as a result of meditation or other means, and had something that we in our ignorance vaguely call a 'mystical experience.' He is profoundly changed by it and wants to share it with the world. He usually indicates that the experience itself is not directly expressible, so he makes some kind of explanation of it, some kind of translation, taking into account the times and place he lives in, the particular culture and people he's been brought up with, so people will be able to understand it. It's extremely powerful for the people of the time.

Then the founder's translation starts getting adapted and further translated by the founder's followers. They, as well as the founder himself, are affected to a strong extent by the social conditioning they've gone through. Over time this powerful experience gets worked into the social structure. Pretty soon the priesthood is part of the governing class, and having proper religious beliefs is conducive to social order. There's a long-term social adaptation, which often produces religious traditions in which primary experience is now a very small part. The morality, the ethics, the understanding from it tend to become are conditioned in people as the result of education, rather than the religion being something that really flows out of their own experience.

There's a funny Sufi story about the sage Nasrudin who was once brought a duck by a cousin in the country. He was very pleased and he made some duck soup and he and the cousin had a nice feed. A few days later someone stopped in who was a friend of the cousin and had heard about the duck. Hospitality required that Nasrudin give him a meal. There was a little duck soup left and they had that. And pretty soon a friend of a friend dropped in, until finally it reached a point where someone who knew the friend who knew the friend who brought the duck was dropping in for hospitality. And he was served also, and he had to remark, 'There is a very delicate texture to this duck soup, it is almost like hot water,' which is what it was. And Nasrudin pronounced this was the soup of a soup of a soup of a soup. I'm afraid it happens that way in a lot of spiritual traditions. What started out as an incredible degree of insight and importance from someone's powerful ASC experience gets progressively diluted and weakened until people no longer know anything about the actual experiential truths that led to the tradition in the first place. They only know the commentaries on the translations of the commentaries of the followers' disciples' commentators.

Something else can happen that's important. The founders of some religions realize that this dilution will probably happen, so they not only work out important insights and values about reality, but they also work out a set of practices, a *technology* as it were, for re-inducing that primary experience so people can experience it for themselves. Buddhism is a primary example of this. Buddha did not intend that somebody should write down what he said so that everybody could simply memorize it. The point is to experience certain kinds of things for oneself. So if you're lucky, in your tradition there is a set of technologies worked out, a set of practices that keeps the experience alive. New people have the experiences and so it stays very alive.

Historically, even so it may happen that after a few generations the understanding becomes diluted. I'm told most Buddhist monks simply memorize scriptures — that it's actually very rare to meditate.

The result of this in the long run is that the original, very powerful experiences that touch people at a very profound level, that join them with others in the universe so that they *know* we are all one, get watered down. It starts looking like superstitions and there is a scientific rejection of the whole practice. That unfortunately happens. But that experience is where deep values come from, and the rejection of the diluted tradition also rejects the primary experiences.

I'm particularly interested in this area then, both because I'm interested in altered states of consciousness and, being interested in human potentials, I want to know about the things that will allow people to get back to the sources of value instead of getting it third or fourth or fiftieth hand.

Now the final question I asked for setting the stage was why was I, as a scientist, interested in transpersonal psychology? To answer that, I must make a confession. I am actually a human being before being a scientist, in spite of all my training. I'm interested in these deep sources of value myself. I'm interested in these insights and practices. They mean something to me personally. I'm particularly interested because in our times much of the turmoil of modern life occurs because we don't have either a value system or a technology by which the majority of people can get back to these deep roots within themselves and actually touch values that are really meaningful.

Secondly, I'm quite aware that science is not practised in a vacuum. You know about textbook science. There's an 'objective observer' who sees the facts as they are and follows them to their logical ends regardless of consequences. And it's total nonsense. Scientists are human beings, they're members of their culture, they share both its virtues and the shortcomings.

Our science, particularly, does not have a very good cultural context to support and guide it. Science in a sense needs a transpersonal, spiritual kind of orientation. It needs something, not as a holy book that says: 'Thou shalt only investigate A, B and C!' but something to remind practising scientists of the consequences of their work and of a wider view of man in which to carry out that work. Furthermore, and perhaps most importantly, I think the spiritual needs science as well. All spiritual disciplines need some way to test the consequences of what they are doing — to more objectively test their concepts — to find out when they're doing the right thing and when they're going astray.

To put that another way, I think there are a lot of spiritual practices around that are no longer appropriate. They were all right at a certain time and place for people of a certain temperament, but they don't work any more. Or they give the wrong kinds of effects. Or they're being used in a context that produces different effects than what they were intended for. The result tends to be institutionalized nonsense that diverts people from really being alive, from really being effective in the world. So, while scient*ism*, the religion that thinks that current equations are revealed truths engraved by God on neutronium or something at the beginning of the universe, may conflict with the spiritual, I don't think real science does.

Science and humility

I want briefly to enlarge on the essence of real science. The real essence of science is humility. Essential science starts with being interested in something and knowing you don't know much about it. You learn about the field you're interested in by carefully observing it. Science institutionalizes humility, as it were, a realization that while you would like to think you are a fantastic observer and nothing ever escapes you, in point of fact you're probably a poor and probably biased observer. Thus it's a good idea to develop systematic techniques of observation and/or instruments of observation to correct for your shortcomings as an observer. That gives you *facts*.

Now it turns out we're actually not too interested in facts. If I told you there are 276 cars parked within three hundred yards of here, would you be excited? So what? We're really interested in what the facts *mean*. People are conceptualizers. For all the virtues of being able to stay just with experience, we love to think about things and *why* they are the way we observed them. In formal terms we theorize. You take the data and you explain why it happened that way. In science you try to be logical about it. Now that's not a simple process because it turns out that logic itself is arbitrary, that there are different kinds of logic, etc. But if you're working with a certain kind of logic like mathematics, you try to follow the rules and you don't say that two plus two equals five if you want to claim to be logical.

In retrospect, though, it turns out that we're too good at thinking. In retrospect we can find a plausible 'reason' for anything. I love to give my classes absolutely ridiculous things to think about, like spilling a can of green paint in the middle of the road at an intersection. Now I tell them to give me four good reasons for doing that. People can come up with such good reasons for doing any random action! One of the things we've learned about the thinking mind, in a sense, is that it can rationalize anything in retrospect. You can always find a framework to apparently explain it.

So good science doesn't stop with coming up with logical theories. It says: 'All right, you thought up a theory. It's mathematical, it's elegant, it draws from the greatest authorities and the most sublime principles and we both love it, *but will it accurately predict what's going to happen next?*' That's the crucial test. If my theory says that this object will not drop because I'm using the magic mantra, 'Gravito, gravitare, gravitato, gravitatus!' and I let go and it drops, it means: I'm sorry, it doesn't matter how elegant my theory and my explanation are, they're not right. It's back to the drawing board, I've got more to learn. That prediction requirement is a way of going back to reality to test out whether your concept, your internal maps, really fit.

There's another aspect of humility that is also relevant. In certain ways you're a flawed thinker, you're a flawed predictor and you admit it. So you talk to others at all stages of this process of essential science, you share your observations and thinking with people. Now they may be a little crazy and flawed too, but on a purely statistical basis they're probably not flawed in exactly the way you are, so they can serve as a check on your particular idiosyncrasies. As you keep recycling this process then of observing, theorizing, testing and predicting, communicating all aspects of it, you start out in great ignorance and move toward a conceptual system where your explanations fit reality better and better. This is essential scientific method.

You may notice I haven't said anything to the effect that essential scientific method should be applied only to the physical world. I think the tragedy of scientific method is that this excellent problem-solving technique has mostly been applied only to the physical sciences.

Science and religion in conflict

Now, where is the conflict then between science and religion? The conflict arises because culturally, this scientific method has been married to a philosophy of materialism that insisted that *all* 'scientific' explanations must be ultimately physical. That marriage is totally unnecessary. That marriage is not the essence of scientific method, that just happens to be an historical artefact. *The scientific method can be applied to the mind. It can be applied to the spirit.* People who have spiritual experiences can try to observe them better, can look at their formulations, can share these processes, can test the consequences, and so forth. For instance, I may come up with the startling insight that love conquers all and nothing can stand in the way of it. And when I meet another person I radiate love to her but she hits me on the head! So I have to realize that my theory is apparently not quite complete ... Perhaps my theory that love conquers all is wrong, or at least there must be some other factors to put in, and that's all right, I learned something from observing that my prediction did not work out.

The traditional science and religion conflict is not really between the essence of science and the essence of spiritual growth. The conflict is between arrogance and humility. Whether you are a scientist or a mystic, if you don't have humility, if you assume that what you already know is absolute truth, then you're in for conflict. The conflict is between second rate science and second rate mystics, between dogmatic people. I've never had any trouble really talking and exchanging views with people who I thought were very spiritually evolved. That's because, although they think things are a certain way, they are certainly willing to think about alternatives. I'm not sure of the way things are, so I'm going to listen to alternatives. When people become psychologically attached to their beliefs, though, then they become defensive, they feel the need to attack other people's beliefs, and conflict arises. So the real distinction then is between arrogance and humility, between closed-mindedness and open-mindedness.

These Mystics and Scientists conferences are very important because they promote open-mindedness. Obviously no one is here who already knows the whole truth. If you already knew all truth, why bother to come and listen to anybody? Everyone has a certain degree of open-mindedness. Once you have a certain amount of open-mindedness, you can apply this basic scientific method to your spiritual life, to both ordinary life experiences and special experiences. You try to be clear on what happened, and you realize you make interpretations of what you've experienced. Do your interpretations make (logical) sense? And do they work? When you apply those interpretations to the rest of your life, do they actually work out, do they provide useful predictions of consequences that can guide your life? If they don't, how can you make better observations? How can you do clearer thinking?

Most important of all, and what I'm afraid is all too lacking in the spiritual community, is *sharing* these kind of experiences and beliefs, data and theories. There's very little sharing of spiritual experiences. Too much of the time, when people with 'spiritual' experiences talk about them to someone else, it's not to openly share and gain others' perspectives, it's to argue, it's to convert. It's to say 'Look, I've got the truth and you'd better listen!' That's not a free flow of information; that's an attack, and other people naturally become defensive.

I think we could have much more progress in the spiritual realm if people could get down to saying, 'What I experienced was this. My interpretation is that. The consequences are this other thing. What did you experience, what do you think? How do your experiences relate to mine?' We would have some real sharing there. That doesn't mean, of course, that people shouldn't learn from their experiences or act on them, but that they should be open-minded about them. My long-term hope, then, is that science will greatly enrich the transpersonal, spiritual world, by bringing in this kind of open-mindedness in and, in turn, that the transpersonal, spiritual world will enrich science.

Transpersonal psychology

Now, let me say a little bit about the content of transpersonal psychology. Transpersonal psychology as discipline is the newest branch of psychology, which probably exists mainly in California. We have a journal, the *Journal of Transpersonal Psychology*, and a professional association, the Association for Transpersonal Psychology.*

Transpersonal psychology has an interest in looking at those aspects of Man that go beyond what are called the first three forces in psychology. The first force is sort of classical behaviourism. Karl Pribram has talked about it nicely as an intra-organism stimulus-response psychology. The second force is classical psychoanalysis, which says that, except for social inhibitions, we'd all be out raping and killing, to terribly over-simplify the basic psychoanalytic message. It's very useful in some ways but not enough. The third force is humanistic psychology, which focuses on more positive human relationships. It teaches that you can have much richer relationships with other people, and live a more fulfilling life, yet there is something still self-centred about it. It still has to do with *you* as an individual and a particular personal history, living at a certain time, having a certain social role, etc.

Transpersonal psychology is really the psychological attempt to begin investigating the spiritual, using psychological insights as well as traditional spiritual insights. Now, the total goal is very grandiose. I'm going to read you the definition of transpersonal psychology that appears in the front of each issue of the *Journal of Transpersonal Psychology:*

* Author's note: Since this talk was given in 1981, considerable expansion and progress has occurred. An outstanding graduate degree program in transpersonal psychology exists at the Institute of Transpersonal Psychology (ITP), 744 San Antonio Road, Palo Alto California 94303, 650 493-4430. Website *http://www.tmn.com/itp* will also lead to information on the Association for Transpersonal Psychology.

Transpersonal Psychology is an interest in those human
capacities and potentialities that have no systematic place in
positivistic, behaviouristic psychology, classical
psychoanalytical theory, or humanistic psychology. It's
concerned with the empirical and scientific study of, and
responsible implementation of, the findings relevant to
becoming, individual and species wide metaneeds, ultimate
values, unitive consciousness, peak experience, being values,
ecstasies, mystical experience, awe, being, self actualization,
essence, bliss, wonder, ultimate meaning, transcendence of the
self, spirit, oneness, cosmic awareness, individual and species
wide synergy, maximal interpersonal encounter, sacralization of
everyday life, transcendental phenomena, cosmic self humour
and playfulness, maximal sensory awareness, and related areas.

Transpersonal psychology itself as a field is mainly a matter of interest
in these things rather than answers.

I'm going to give an example of what's typically been called a
transpersonal experience and talk about some aspects of it. I particularly want to talk about the problems we have in making a living
reality out of the transpersonal, as opposed to merely inspiring ideas.
I sometimes use this case as an example of an out of the body
experience, but it really has quite important transpersonal aspects. It
was recorded by Lord Geddes many years ago, about an anonymous
physician friend:

It was on September 9th, a few minutes after midnight, I began
to feet very ill, and by two o'clock was definitely suffering
from acute gastroenteritis which kept me vomiting and purging
until almost eight o'clock. By ten o'clock I developed all the
symptoms of acute poisoning, intense gastro-intestinal pain,
diarrhoea, and pulse and respiration so high that I couldn't
count them. I wanted to ring for assistance but found that I
could not, so quite placidly gave up the attempt. I realized I
was very ill and very quickly reviewed my whole financial
position.

[That's just a little guarantee that I'm talking about a sane,
solid, respectable citizen here, not some weird, fly-by-night
person.]

Thereafter, at no time did my consciousness appear to be in any way dimmed, but I suddenly realized that my consciousness was separating from another consciousness which was also me. These for purposes of description we could call the A and B consciousness. Throughout what follows ego attached itself to the A consciousness. The B personality I recognized as belonging to my body, and as my physical condition grew worse, with my heart fibrillating rather than beating, I realized that the B consciousness belonging to the body was beginning to show signs of being composite, that is built up of consciousnesses in the head, the heart and the viscera. These components became more and more individualized, and the B consciousness began to disintegrate, while the A consciousness which was now me, seemed to be altogether outside my body which it could see from outside. Gradually I realized that I could see not only my body in the bed in which it was, but everything in the whole house and garden; and then realized that I was not only seeing things at home but in London, in Scotland, in fact wherever my attention was directed. It seemed to me the explanation I received, from what source I do not know, but which I found myself calling my mentor, was that I was free in a time dimension of space where 'now' was in some way equivalent to 'here' in the ordinary three-dimensional space of everyday life.

The narrator then says his further experiences can only be described metaphorically. So, while he seemed to have ordinary two-eyed vision, he 'appreciated' rather than 'saw' things. He began to recognize people he knew in his visions and they seemed to be characterized by coloured condensations around them. He goes on to report:

Just as I began to grasp all these I saw my wife enter the bedroom. I realized she got a terrible shock and saw her hurry to the telephone. I saw my doctor leave his patients and come very quickly. And I heard him say, or rather I saw him think, 'He's nearly gone.' I heard him quite clearly speaking to me on the bed, but as I was not in touch with my body, I could not answer him. I was really cross when he took a syringe and rapidly injected my body with something which I afterwards

learned was camphor. My heart began to beat more strongly, I
was drawn back, and I was intensely annoyed because I was so
interested and just beginning to understand where I was and
what I was doing. I came back into the body really angry at
being pulled back and all the clarity of vision left me.

This is a classical kind of transpersonal experience in a variety of
ways. Firstly, a basic instinct was altered. We think survival of the
physical body is the ultimate instinct, but he quite placidly gave it up
after a point. It didn't matter to him. He recognized the mixed character of some aspects of his consciousness. He recognized how much his
body contributed to what he thought was ordinarily his consciousness,
and recognized that he had a certain consciousness other than his body
itself. He had an out of the body experience, including seeing his body
from the outside; he learned that the primary thing in consciousness
was his attention or will. He could see wherever he wished to see; he
wasn't limited by anything. There was a transcending of spatial limits
that he experienced, a psychic transcendence of some kind or another.
He experienced forces of wisdom, the things he called his mentor,
other kinds of entities or beings. He experienced a new relationship in
time and space and a new way of perceiving things; for instance, he
saw auras around people. But his understanding was what we now
technically call *state-specific*. That is, when he was in this state of
almost being dead, his understanding made perfect sense, but it did not
transfer back to his ordinary state. He could only use metaphors and
could not really grasp what it was, but he understood it at that time.
This state-specificity is actually quite common in transpersonal experiences. And, of course, I think we can presume he had a rather deep value
change. His attitude toward death certainly changed quite drastically.

Now this type of experience has 'come out of the closet' recently
and experiencers have started talking about it, whereas a few years ago
no one mentioned it. Raymond Moody published a book called *Life
After Life*, based on his interviews with people who had clinically died
and then been resuscitated. Now to avoid a long semantic argument,
they weren't, of course, *really* dead. Since death is a permanent state,
they were not really dead; but let's say these were people who came
as close as they can get and still tell us anything about it. Many of
them were pronounced dead by their doctors.

Moody puts together what he calls the common features of all these cases. Some of these are: hearing oneself pronounced dead (often a quite shocking experience, however sometimes quite a relief, depending on how much suffering the person was having before); some kind of buzzing or vibration sensation; frequently a feeling of going through some kind of tunnel toward a light at the end; finding oneself outside one's physical body; looking at it from outside; watching the resuscitation attempts. There are often parapsychological aspects here, such as the persons correctly describing things about resuscitation attempts that they should not have known from their non-conscious position inside the body.

People usually find themselves in some kind of second non-physical body; and often meet other non-physical entities, spirits of friends or relatives, who seem to be there to help them make the transition. They often meet what's most neutrally called a *being of light*. Cultural effects may show up here. Christians say this being was 'Christ,' Jews say it was an 'angel,' atheists, not wanting to commit themselves theologically, just say a being of light. This luminous being literally radiates extremely bright 'light,' except it's a funny kind of light — it doesn't hurt the eyes. Experiencers frequently undertake a review of their life in which the highlights of their life flash back before them. Sometimes they insist that they see not just the highlights but *every single event* in their life. They judge themselves, which is far worse than being judged by someone else, because you can't kid yourself very well in this state.

They often see some kind of barrier or border in this state, and they know that if they cross it, they'll be permanently dead. Now this is a methodological artefact of our case collection method, of course, but none of the people we hear from have crossed this particular barrier. They've all come back, sometimes quite angrily because they didn't want to return; they were usually experiencing tremendous joy and peace and serenity in this other state and quite often they don't want to come back. At other times they may say, in spite of how wonderful it is, they've got responsibilities here; there are things they haven't finished, and they come back voluntarily.

They usually say the experience is *ineffable* in the sense that they try to describe it in words, but it doesn't really convey the experience; there are too many things that they just don't have words for. They may try to tell some people about the experience and usually they rather quickly find that they're written off as a nut. If it happens in a hospital they're liable to send the psychiatrist around, because the patient's obviously out of his head. So they stop telling people about the experience.

Their lives usually change dramatically as a result of the near death experience, especially the life review part. In the life review, they see what's really important in life. They may say, 'I finally learned that I never learned to be like a child, and I've got to stop wasting my time on trivial things in life and learn that.' Or, 'I've never learned to love, that's what's important to learn.' One of the best ways to have your values profoundly changed is almost to die. But I don't recommend the technique because the 'almost' part is very tricky. While it's fast, it's not a recommended technique!

And, of course, their ideas about death change drastically. They usually say something like: 'I no longer *believe* in survival of death. I *know* I'll survive death. It's not a matter of belief, I've been there.' Now again, those of us who have not had that experience can argue about that, we can say 'Your body was alive so you weren't really dead,' and they say, 'Those are just words, but I *know*.'

You don't have to almost die, though, to have some deep value changes. There are spiritual techniques that can be practised, sometimes for a short time, sometimes for very long times, that can produce altered states of consciousness and profound experiences. transpersonal psychology is even beginning to produce some of these experiences in the laboratory and study them. For example, I conducted an hypnosis experiment exploring very deep hypnosis, much deeper than hypnosis is ordinarily taken, and outside of the usual sort of authoritarian context where I am the hypnotist and you are in my power, much more a relationship between equals-helping someone explore something. A young man whom I've nicknamed William was rating how deeply hypnotized he was on a self-report scale of hypnotic depth. Most people would never go beyond, say 15 or 20 on this scale; 30 would be considered deep hypnosis, but he could go to 50. William experienced various kinds of phenomena in our explorations, and I would ask him about their relative intensity as a crude way of initially mapping this. So, for instance, he initially got more and more physically relaxed, and at a certain point if I asked him about relaxation, he couldn't tell me how relaxed he was. His body was left behind, it was just a feeling. Things were neither relaxed or tense; he simply couldn't rate that kind of thing.

He had a sensation of visual blackness as he went deeper that became 'absolutely black' and then got even 'blacker.' But it was 'filled up' with something even though there was nothing there to see and it was totally black. I asked him about feelings of peacefulness, as people often feel peaceful while hypnotized. He got extremely peaceful, but then again when he reached a certain profound depth, peacefulness simply wasn't a concept that made any sense. He couldn't rate how peaceful he was, he was beyond that particular concept.

William's awareness of the physical environment around him faded out to basically nothing, fortunately with the exception of my voice. I asked him who he was, and what his sense of identity was. That sense was initially centred in his head, but this feeling of 'Yes, I'm me,' gradually faded out and became more transpersonal. Who he was became *potential*. He could be anybody after a certain point — it made no sense to talk about his identity in terms of his ordinary self, he simply became something else entirely.

What else did he experience? One experience we called 'awareness of The Joke.' How to explain The Joke? If, when you have a sort of cosmic humour about the fact that you think you're unenlightened and you're trying to do something about it, it's really quite funny, not in a put down kind of way, but it really is quite amusing that you go through all these kinds of antics to get enlightened because you really are already enlightened. Don't get it? We're not in the right state of consciousness to get The Joke, I'm afraid!

During these explorations, who was I, the experimenter? William was very focused on me in ordinary hypnosis for a long way. Then I began to seem more and more distant, a remote kind of 'thing.' And then he said I became just an amusing tiny ripple on the far fringes of the sea of consciousness, a ripple who would ask him silly things like what he was experiencing now. His experience of time passage changed also. Time seemed to go slower up to a point but then he simply thought he was no longer *in* time — time itself was a meaningless concept. How could he talk about that kind of thing? His feeling of oneness with the universe increased steadily, his spontaneous mental activity went down so basically he would think of nothing, his mind was not going chatter, chatter, remember, think, plan, etc., etc. as we ordinarily do. The one other phenomena he experienced was a change in his awareness of his breathing, he fantasized it becoming deeper and deeper and then it just disappeared. Breathing was no longer relevant

Now this is not a typical example of transpersonal psychology research, but it's an example of a transpersonal experience happening in the laboratory in a way that you can begin exploring it. This does not mean that you can run out to your corner hypnotist and definitely experience this. William was an especially talented young man and this was quite profound work we had been doing for a long while, but it's an example of the kind of contribution science can begin to make. We can look at this type of experience; it's not taboo.

A wider view of the human being

Now, let me start putting my previous remarks together into a wider view of the human being, a view that's my personal understanding, influenced by the kinds of human experiences that I talked about plus many more. I think this view is compatible with science in the long run, but certainly it's not a clear scientific theory at this point. I have done formal theorizing on this in some of my scientific writings, but I'm not going to limit myself to that.

The view I have is that firstly, there is a real aspect of a human being that we call 'mind,' for want of a better word, that's not reducible to the physical brain in principle. Psychic data require this, experiences in altered states and mystical experiences suggest this. Look at the consequences of this first point.

For those of you who are philosophically inclined, I am labelling myself as a *dualist*. But I'm a special kind of dualist — I'm a *pragmatic dualist*. Most philosophical arguments about monism and dualism frankly bore me, because they have no real, observable consequences. It's not clear how anything would be different in the world if one or the other view were true. But mine is a pragmatic dualism because what I really say is that I don't care so much about absolute differences. I don't know if mind and body are absolutely different in some sense. I stress that they are different in sense that has methodological consequences, namely you can't simply wait for advances in the physical sciences to explain all mental phenomena. Mystical, spiritual, and parapsychological phenomena have to be investigated on their own and we have to develop methods and understandings peculiar and specific to those kind of phenomena in order to understand them properly. So that's the point of this kind of pragmatic dualism.

The fact that the mind can seem to transcend space and time on occasion suggests to me that there are real realms of experience in which the mind is, as it were, not very clearly localized. It's 'spread out' somehow through time and through space, and this spread is manifested as psychic abilities of some sort. It would be hard to say *where* a mind was or *when* a mind was or exactly what its nature is when this happens. I think it is potentially possible to understand it, but we can't specify it very clearly at this time.

When mind and body interact, *psi* phenomena, (such as telepathy, clairvoyance and psychokinesis) are the mechanisms, the processes by which they interact. There are two components: a body/brain nervous system component and this mind component. They interact with each other through clairvoyance and psychokinesis.

However, there are large parts of the brain that are probably autonomous. They don't need any mind, and psychic interaction with mind *per se*, to operate. They run on their own, the system's on automatic. But there are occasions when the physical brain interacts with this nonphysical part of mind. The mind uses clairvoyance to know what the physical state of the brain is and thus receive information about the environment that the brain has picked up through the senses, and the mind uses psychokinesis to affect the firing of the brain. This psychokinetic type of interaction will probably turn out to be in areas of the brain that are almost like random generators, neurons are almost ready to fire and instigate certain kinds of patterns, and some very small influences can set them going. Thus in my view we use psychic abilities all the time, but *inside* ourselves, as it were. It's the manifestations of psi outside ourselves that are the rare and unusual ones.

Knowing that reality is undoubtedly complicated, it's likely that both brain and mind are organized on a multitude of levels, and they may interact psychically in a multitude of levels. The result of all this interaction is *consciousness as we experience it.* That is, when you describe your consciousness, when you look inside to see what you're like, you're not experiencing just your brain and you're not experiencing just your mind. You're experiencing what is technically called a *systems emergent*, a gestalt resultant of these two components, mind and brain, interacting in various ways. As I said, psi is ordinarily used 'inside' the organism you think of as yourself, this compound of mind and brain. This use is what I've named *auto-clairvoyance* and *auto-psychokinesis* for the mind-brain link. On rare occasions psi is used to obtain information directly from the outside, physical world, what I've named *allo-clairvoyance*, or to directly affect the outside world, *allo-psychokinesis*. Indeed it's these rare manifestations of psi 'outside' our own organisms that lets us know that psi exists and gives us some clue as to what it is.

The result of this mind-brain interaction is to localize our consciousness in time and space. That is, the mind is becoming embodied as it were, interacting psychically very strongly with a particular brain and nervous system and body, producing consciousness as we experience it. It also produces a physical here and now which it is adaptive to be very aware of, as biological survival and gratification are rather important to us.

From this perspective, allo-psychic abilities, direct psychic perception of the external world, are, in a sense, relatively useless. Or let's say they're not very useful for everyday living. What's biologically relevant most of the time is what's immediately near you, and our ordinary physical senses are usually superb for telling us what is physically nearby. If you're going to cross the street, it's much better to use your eyes and ears to see that truck coming around the corner than to psychically know what's going on in a distant place at that time! The distant place isn't quite as relevant as the truck.

A science of the spirit

There's one other theme I want to mention. Unless these transpersonal ideas and spiritual ideas in general, are integrated in a practical way with good quality scientific knowledge about the world, they are not only useless, they're dangerous! To simply go around and tell people they have wonderful, positive spiritual potentials can often be a disservice. Because unless you know how to realize them, what to do about them, they lead to useless daydreaming and fantasy life. For instance, for most people I find the idea of reincarnation useless, because most of the people I've met who are interested in reincarnation are making a terrible mess of this life. But instead of cleaning up their act and getting it together, they think, 'Oh, but I was so wonderful last time, and have such a wonderful future life ahead of me!' These ideas, unless they lead to practical application, aren't really very helpful.

I want to say just a few things about why we don't develop our transpersonal potentials. Why don't we simply become enlightened and develop our psychic abilities and realize our full human potential? What is the barrier? Spiritual traditions in general point the way to, I think, a real understanding of some of the obstacles. For instance, you have Eastern ideas of *samsara* or *maya*, the idea that the world is illusion, which in my understanding is not that the world *per se* is unreal, it's that we live in such a fantasy world that we stumble around in the actual world rather blindly. That concept is fine as far as it goes, but what about the details of living in illusion, so that you can actually navigate and develop more effectively in this world.

Most of my work is on the psychology of consciousness. I'm interested in our ordinary mind as well as special altered states. We're born as human beings with an enormous number of potentials. You have the potential to speak fluent Chinese. You have the potential to go into trances where friendly spirits will possess you and teach you beautiful songs and dances that you can share with people. You have the potential for really esoteric skills and knowledge, like, e.g., integral calculus or quantum physics.

But you're a member of a particular culture, and a culture is a group of people who knows about certain human potentials and deliberately chooses some of them, ignores others, and actively rejects still others. That is, out of all the things you could become, culture already starts limiting you in the educational process. Your parents usually act as the main agency of enculturation. Your parents start telling you what's wrong and what's right, how you should think, how you should feel, how you should act. Even in infancy, teachers and peers begin to have a certain amount of influence to further tell you the way things are. By the end of your childhood you have become a basic member of your culture. Your consciousness has been constructed and shaped to reflect the assumptions of your culture. By late childhood you are simultaneously, as Aldous Huxley so nicely put it, simultaneously the beneficiary and the victim of your culture. By adolescence, parents and teachers disappear as significant influences as we all know, and only what their peers want is important. By adulthood, your peers in a sense have been internalized. You don't need somebody telling you the rules all the time. It's become a habitual part of your thinking. By old age you either have an opportunity to cut loose and really understand yourself, or you may just accept social definitions of your uselessness and pine away.

Our perception, then, is never a simple realistic perception of what the world is. The point is that there is a lot going on between you and your perceptions, and much of it you don't realize. Some of it you can't do much about, such as the way your eyes are built, and except for instrumental aids, you're stuck with that. The cultural conditioning you can do something about, though.

Now, I mentioned earlier when I explained scientific method that it was also capable of being a marvellous spiritual seeking. Scientific method is refined common sense: if you're interested in something, you try to see what's there as best you can and always improve your observations; you try to think about it and make sense out of it and to test the consequences of understanding, and share things with people so you get the benefit of their view. However, cultural conditioning perverts the scientific process, perverts clear thinking. Instead of potentially being interested in *everything*, the culture immediately narrows it down to acceptable fields of interest. The kind of thinking you do about it is *consensus thinking*. You want to be logical, sensible, so that people will respect you. There are certain thoughts you don't have. Your testing of things is actually limited to certain approved actions, so you get a kind of pseudo-validation because you really set things up that way to begin with. And instead of a free flow of information where other people could compensate for your own biases, information exchange becomes very much a matter of seeking approval, of wanting other people to think you're smart, to think you're proper, to think you're clever. The whole process of learning from experience becomes severely restrictive.

What I'm saying then is that there is considerable resistance to realizing our full potential. Our enculturation, our particular history has made certain things very dear to us, has made us afraid of a lot of things, and much of that conditioning is not even conscious. We're all afraid of things we don't know we're afraid of. We're all desperately wanting things we don't know that we want because it's become a habit. It takes special techniques like meditation, various kinds of insight psychotherapies and the like to begin to discover what these unconscious things are.

I mentioned near-death experiences before, and that people have a hard time talking about them, and often stop telling them, particularly if these experiences happen in a hospital. Some of the hardest people to tell them to are doctors. Doctors are chained to the scientific tradition where such experiences are not allowed to happen or must be immediately dismissed as craziness. But who are the hardest people of all to tell these near-death experiences to, the ones who most resist hearing about them? Priests and ministers. Priests and ministers do not want to hear about spiritual experiences. It might not be a *proper* spiritual experience.

This resistance is widespread. We're interested in these areas, but we have a lot of cultural conditioning to overcome as part of really being able to realize those human potentials. Transpersonal psychology as a field of psychology is just developing. It's going to draw from the scientific traditions; it's going to draw from the world's great spiritual traditions; it's going to make some marvellous mistakes along the way; it's probably going to discover some new things and learn about things which do and don't work, etc. But it's still in a very early stage.

The reality I'm dealing with and sharing some of with you today is not a reality full of answers yet, so much as a reality full of exciting possibilities for human beings and full of questions that collectively and individually we have to solve. But it's about our fundamental spiritual, as well as biological nature as humans, and it's better to make mistakes in trying to learn more about it than to sweep it under the rug.

The Mystery of the Human Psyche

SIR JOHN C. ECCLES, FRS

After gaining first class honours M.B., B.S. from Melbourne University, Sir John Eccles studied at Oxford where he was awarded first class honours in natural Science and gained his M.A. and D.Phil. He became Professor of Physiology at the University of Otago and then at the Australian National University and was President of the Australian Academy of Science 1957-61. Later, he joined the Institute for Biomedical Research in Chicago before becoming Head of the Research Unit of Neurobiology at the State University of New York at Buffalo where he was Professor Emeritus. He was made Honorary Fellow of Exeter College, Oxford in 1961, Magdalen College, Oxford in 1964, and the American College of Physicians in 1967. He has been awarded many other Honorary Doctorates and distinguished awards. In 1963 he won a Nobel Prize in Medicine. His recent publications include: The Understanding of the Brain; The Self and its Brain *(with Karl Popper);* The Human Psyche *and* The Human Mystery; The Evolution of the Brain; Creation of the Self *and* How the Self Controls its Brain. *He died in 1997. This lecture was delivered at the 1982 Mystics and Scientists conference.*

It is proposed to use the term *self-conscious mind* for the highest mental experiences of the psyche. It implies knowing that one knows, which is, of course, initially, a subjective or introspective criterion. However, by linguistic communication it can be authenticated that other human beings share in this experience of self-knowing. One has only to listen to ordinary conversation, which is largely devoted to recounting the conscious experiences of the speakers. At a lower level, there can be consciousness or awareness as indicated by intelligent learned behaviour and by emotional reactions. We can speak of an animal as conscious when it is capable of assessing the complexities of its present situation in the light of past experience and so is able to arrive at an appropriate course of action that is more than a stereotyped instinctive response. In this way it can exhibit an original behaviour pattern which can be learnt, and which includes a wealth of emotional reactions. It will be appreciated that there is no indubitable test for consciousness, but it is generally accepted that birds and mammals display conscious behaviour when they act intelligently and emotionally and are able to learn appropriate reactions.

The human person

Each of us continually has the experience of being a person with a self-consciousness, not just conscious but knowing that you know. In defining 'person,' I will quote two admirable statements by Immanuel Kant: 'A person is a subject who is responsible for his actions'; and 'A person is something that is conscious at different times of the numerical identity of its self.' These statements are minimal and basic, and they could be enormously expanded. For example, Karl Popper and I have published a 600-page book on *The Self and its Brain*. In it Popper refers to 'that greatest of miracles: the human consciousness of self.' (1977, p.144)

We are apt to regard the person as identical with the ensemble of face, body, limbs, and so on, that constitute each of us. It is easy to show that this is a mistake. Amputation of limbs, losses of eyes, for example, though crippling, leave the human person with its essential identity. This is also the case with the removal of internal organs. Many can be excised in the whole or in part. The human person survives unchanged after kidney transplants or even heart transplants. You may ask what happens with brain transplants. Mercifully, this is not feasible surgically but even now it would be possible successfully to accomplish a head transplant. Who can doubt that the person 'owning' the transplanted head would now 'own' the acquired body, and not vice versa! We can hope that with human persons this will remain a Gedanken experiment, but it has already been successfully done in mammals. We can recognize that all structures of the head extraneous to the brain are not involved in this transplanted ownership.

For example, eyes, nose, jaws, scalp, and so on, are no more concerned than are other parts of the body. So we can conclude that it is the brain and the brain alone that provides the material basis of our personhood.

But when we come to consider the brain as the seat of the conscious personhood, we can also recognize that large parts of the brain are not essential. For example, removal of the cerebellum gravely incapacitates movement, but the person is not otherwise affected. It is quite different with the main part of the brain, the cerebral hemispheres. They are very intimately related to the consciousness of the person, but not equally. In 95% of persons there is dominance of the left hemisphere, which is the speaking hemisphere. Except in infants, its removal results in a most severe destruction of the human person, but not annihilation. On the other hand, removal of the minor hemisphere (usually the right) is attended with loss of movement on the left side (hemiplegia) and blindness on the left side (hemianopia) but the person is otherwise not gravely disturbed. Damage to other parts of the brain can also greatly disturb the human personhood, possibly by the removal of the neural inputs that normally generate the necessary background activity of the cerebral hemispheres.

So to sum up the evidence, we can say that the human person is intimately associated with its brain, probably exclusively with the cerebral hemispheres, and is not at all directly associated with all the remainder of its body. The association that you experience of limbs, face, eyes, and so on, is dependent on the communication by nerve pathways to the brain, where the experience is generated. We are on the threshold of brain-mind problems that will be fully considered in a later section. Our immediate concern is with the question: how does a human person come to exist?

Ontogeny of the human person with self-consciousness

Let us now briefly consider how a human embryo and baby becomes eventually a human person. It is a route that all of us have traversed, but much is unremembered. A baby is born with a brain that is very fully formed in all its detailed structure, but of course it has yet to grow to the full adult size of about 1.4 kg. The nerve cells, the unitary components of the brain, have almost all been made. All the major lines of communication from the periphery and from one part of the brain to another have been grown ready for use. Much before, the brain has been causing the movements sensed by the mother. And even before birth the child can respond to sounds. Its hearing system is already functioning well by birth, which is far earlier than vision. It is remarkable that by seven days after birth a baby has learnt to distinguish its mother's voice from other voices, just as happens with lambs. Then follows a long period of learning to see and to move in a controlled manner.

As we all know, even in the first months of life a baby is continually practising its vocal organs and so is beginning to learn this most complex of all motor co-ordinations. Movements of larynx, palate, tongue, lips, have to be co-ordinated and blended with respiratory movements. It is another variety of motor learning, but now the feedback is from hearing, not from vision. Vocal learning is guided by hearing and is at first imitative of sounds heard. This leads on to the simplest types of words like dada, papa, mama, that are produced at about one year. It is important to realize that speech is dependent on feedback from hearing the spoken words. The deaf are mute. In linguistic development recognition outstrips expression. The child has a veritable word-hunger, asking for names and practising incessantly, even when alone. It dares to make mistakes devolving from its own rules, as, for example, with the irregular plural of nouns. Language does not come about by simple imitation. The child abstracts regularities and relations from what it hears and applies these principles in building up its linguistic expressions.

The investigations of Amsterdam led him to give eighteen months for the transition from the conscious baby to the self-conscious child. He used the same technique as Gallup did, but with the red mark on the face of the child. The children's reactions showed that they recognized the mirror image as their own.

To be able to speak, given even minimal exposure to speech, is part of our biological heritage. This endowment has a genetic foundation, but one cannot speak of genes for language. On the other hand the genes do provide the instructions for the building of the special areas of the cerebral cortex concerned with language, as well as all the subsidiary structures concerned in vocalization.

The earliest stages of functional development may be almost entirely pragmatic, as the child uses its protolanguage to regulate those around it, to acquire desirables, and to invite interaction. These protofunctions in which the child uses objects as foci for interaction develop into the more mature mathetic function, in which the child uses language to

WORLD 2   WORLD 3

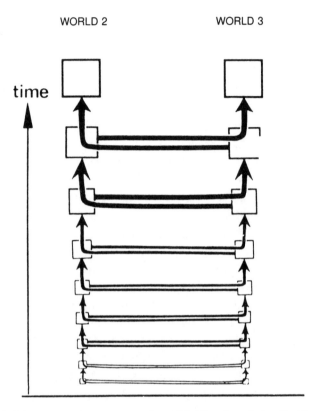

Figure 1. Diagrammatic representation of the developments of self-conscious-ness (World 2) and culture (World 3) in time.

learn about the world — its cognitive aspect. But, of course, these two functions, the pragmatic and the mathetic, are inextricably mixed in the language that a child uses from moment to moment. I would suggest that the remarkable linguistic progress by the child in the first few years is accountable to the developing self-consciousness of the child in its struggle for self-realization and self-expression. Its mental development and its linguistic development are in reciprocal positive interaction.

The Three World philosophy of Popper forms the basis of my further exploration of the way in which a human baby becomes a human person. All the material world, including even human brains, is in the matter-energy World 1. World 2 is the world of all conscious experiences and World 3 is the world of culture, including especially language. At birth the human baby has a human brain, but its World

2 experiences are very rudimentary and World 3 is unknown to it. It, and even a human embryo, must be regarded as human beings, but not as human persons.

The emergence and development of self-consciousness (World 2) by continued interaction with World 3 is an utterly mysterious process. It can be likened to a double structure that ascends and grows by the effective cross-linkage.

In Figure 1, the vertical arrow shows the passage of time from the earliest experiences of the child up to the full human development. From each World 2 position an arrow leads through the World 3 at that level up to a higher, larger level which illustrates symbolically a growth in the culture of that individual. Reciprocally the World 3 resources of the self act back to give a higher, expanded level of consciousness of that self (World 2). And so each of us has developed progressively in self creation. The more the World 3 resources of the individual, the more does it gain in the self-consciousness of World 2. What we are is dependent on the World 3 that we have been immersed in and how effectively we have utilized our opportunities to make the most of our brain potentialities. The brain is necessary but not sufficient for World 2 existence and experience. This is indicated in Figure 2, which is a dualist-interactionist diagram showing by arrows the flow of information across the interface between the brain in World 1 and the conscious self in World 2.

There is a recent tragic case illustrative of Figure 1. A child, Genie, was deprived of all World 3 influence of her psychotic father. She was penned in isolation in a small room, never spoken to and minimally serviced from the age of 20 months up to 13 years and 8 months. On release from this terrible deprivation she was, of course, a human being, but not a human person. She was at the bottom rung of the ladder in Figure 1. Since then, with the dedicated help by Dr Susan Curtiss, she has been slowly climbing up that ladder of personhood for the last eight years. The linguistic deprivation seriously damaged her left hemisphere, but the right hemisphere stands in for a much depleted language performance. Yet, despite this terrible delayed immersion in World 3, Genie has become a human person with self-consciousness, emotions and excellent performances in manual dexterity and in visual recognition. We can recognize the necessity of World 3 for the development of the human person. The brain is built by genetic instructions (that is Nature), but development to human personhood is

BRAIN⇌MIND INTERACTION

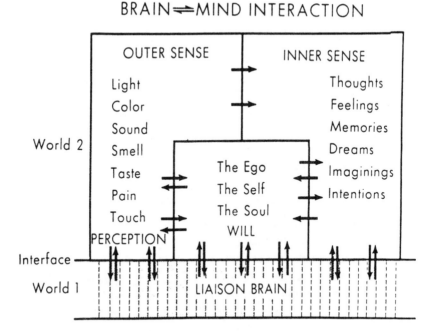

Figure 2. Information flow diagram for brain-mind interaction. Communication lines are shown by arrows.

dependent on the World 3 environment (that is Nurture). With Genie there was a gap of over thirteen years between Nature and Nurture.

It may seem that a complete explanation of the development of the human person can be given in terms of the human brain. It is built anatomically by genetic instructions and subsequently developed functionally by learning from the environmental influences. A purely materialist explanation would seem to suffice with the conscious experiences as derivative from brain functioning. However, it is a mistake to think that the brain does everything and that our conscious experiences are simply a reflection of brain activities, which is a common philosophical view. If that were so, our conscious selves would be no more than passive spectators of the performances carried out by the neuronal machinery of the brain. Our beliefs that we can really make decisions and that we have some control over our actions would be nothing but illusions. There are, of course, all sorts of subtle cover-ups by philosophers from such a stark exposition, but they do not face up to the issue. In fact, all people, even materialist philosophers, behave as if they had at least some responsibility for their own actions. It seems that their philosophy is for 'the other people, not for themselves,' as Schopenhauer wittily stated.

These considerations lead me to the alternative hypothesis of dualist-interactionism, that has been expanded at length in the book *The Self and its Brain*. It is really the commonsense view, namely that we are a combination of two things or entities: our brains on the one hand, and our conscious selves on the other. The self is central to the totality of our conscious experiences as persons through our whole waking life. We link it in memory from our earliest conscious experiences. It lapses during sleep, except for dreams, and it recovers for the next day by the continuity of memory. But for memory we, as experiencing persons, would not exist. Thus we have the extraordinary problem that was first recognized by Descartes: how can the conscious mind and the brain interact?

In contrast to these materialist or parallelist theories are the *dualist-interactionist* theories. The essential feature of these theories is that mind and brain are independent entities, the brain being in World 1 and the mind in World 2, and that they somehow interact, as illustrated by the arrows in Figure 2. Thus there is a frontier, as diagrammed in Figure 2, and across this frontier there is interaction in both directions, which can be conceived as a flow of information, not of energy. Thus we have the extraordinary doctrine that the world of matter-energy (World 1) is not completely sealed, which is a fundamental tenet of physics, but that there are small 'apertures' in what is otherwise the completely closed World 1. On the contrary, as we have seen, closedness of World 1 has been safeguarded with great ingenuity in all materialist theories of the mind. Yet I shall now argue that this is not their strength, but instead their fatal weakness (cf. Popper and Eccles, 1977).

Critical evaluation of brain-mind hypotheses

Great display is made by all varieties of materialists that their brain-mind theory is in accord with natural law as it now stands. However, this claim is invalidated by two most weighty considerations.

Firstly, nowhere in the laws of physics or in the laws of the derivative sciences, chemistry and biology, is there any reference to consciousness or mind. Regardless of the complexity of electrical, chemical or biological machinery, there is no statement in the 'natural laws' that there is an emergence of this strange non-material entity, consciousness or mind. This is not to affirm that consciousness does not emerge in the evolutionary process, but merely to state that its emergence is not reconcilable with the natural laws as at present understood. For example, such laws do not allow any statement that consciousness emerges at a specified level of complexity of systems, which is gratuitously assumed by all materialists except panpsychists. Their belief that some primordial consciousness attaches to all matter, presumably even to atoms and subatomic particles finds no support whatsoever in physics. One can also recall the poignant questions by computer-lovers. At what stage of complexity and performance can we agree to endow them with consciousness? Mercifully this emotionally charged question need not be answered. You can do what you like to computers without qualms of being cruel!

Secondly, all materialist theories of the mind are in conflict with biological evolution. Since they all (panpsychism, epiphenomenalism and the identity theory) assert the causal ineffectiveness of conscious-ness *per se,* they fail completely to account for the biological evolution of consciousness, which is an undeniable fact. There is firstly its emergence and then its progressive development with the growing complexity of the brain. In accord with evolutionary theory only those structures and processes that significantly aid in survival are developed in natural selection. If consciousness is causally impotent, its develop-ment cannot be accounted for by evolutionary theory. According to biological evolution mental states and consciousness could have evolved and developed *only if they were causally effective* in bringing about changes in neural happenings in the brain with the consequent changes in behaviour. That can occur only if the neural machinery of the brain is open to influences from the mental events of the world of conscious experiences, which is the basic postulate of dualist-inter-actionist theory.

Finally the most telling criticism of all materialist theories of the mind is against its key postulate that the happenings in the neural machinery of the brain provide *a necessary and sufficient explanation of the totality both of the performance and of the conscious experience of a human being.* For example, the willing of a voluntary movement is regarded as being *completely determined* by events in the neural machinery of the brain, as also are all other cognitive experiences. But, as Karl Popper states in his Compton Lecture:

> According to determinism, any such theory such as, say, determinism is held because of a certain physical structure of the holder — perhaps of his brain. Accordingly, we are deceiving ourselves and are physically so determined as to deceive ourselves whenever we believe that there are such things as arguments or reasons which make us accept determinism. In other words, physical determinism is a theory which, if it is true, is unarguable since it must explain all our reactions, including what appear to *us as beliefs based on* argument, as due to purely physical conditions. Purely physical conditions, including our physical environment, make us say or accept whatever we say or accept. (Popper 1972, Chapter 6)

This is an effective *reductio ad absurdum*. This stricture applies to all of the materialist theories. So perforce we turn to dualist-interactionist explanations of the brain-mind problem, despite the extraordinary requirement that there be effective communication in both directions across the frontier shown in Figure 2.

Necessarily the dualist-interactionist theory is in conflict with present natural laws, and so is in the same 'unlawful' position as the materialist theories of the mind. The differences are that this conflict has always been admitted and that the neural machinery of the brain is assumed to operate in strict accordance to natural laws except for its openness to World 2 influences.

Moreover, as stated by Popper (1977, Dialogue xii), the interaction across the frontier in Figure 2 need not be in conflict with the first law of thermodynamics. The flow of information into the modules could be effected by a balanced increase and decrease of energy at different but adjacent micro-sites, so that there is no net energy change in the brain. The first law at this level may be valid only statistically.

It is useful to think of the brain as an instrument, our computer, that has been our lifelong servant and companion. It provides us, as programmers, with the lines of communication from and to the material world (World 1) which comprises both our bodies and the external world. It does this by receiving information by the immense sensory system of millions of nerve fibres that fire impulses into the brain where it is processed into the coded patterns of information that we read out from moment to moment in deriving all our experiences — our percepts, thoughts, ideas, memories. But we, as experiencing persons, do not slavishly accept all that is provided for us by our computer, the neuronal machine of our sensory system and of our brain. We select from all that is given according to interest and attention, and we modify the actions of the neuronal machinery of our computer, for example, to initiate some willed movement or in order to recall a memory or to concentrate our attention.

Uniqueness of the psyche

It is not in doubt that each human person recognizes its own unique-ness, and this is accepted as the basis of social life and of law. When we enquire into the grounds for this belief, the evidence just presented eliminates an explanation in terms of the body. There are two acceptable alternatives, the brain or the psyche. Materialists must subscribe to the former, but dualist-interactionists have to regard the psyche or World 2 (cf. Figure 2) as being the vehicle for the unique-ness of the experienced self. This, of course, is not to deny that each brain also is unique. A structure of such complexity can *never* be duplicated down to the most minute detail, and it is the minute detail that counts. This is even true when there is the genetic identity of uniovular twins. But our enquiry concerns the experienced uniqueness, and this can be preserved even when there are profound changes in the brain in surgery, in cerebral injuries and degenerations.

As a dualist-interactionist I believe that my experienced uniqueness lies not in the uniqueness of my brain, but in my psyche. It is built up from the tissue of memories of the most intimate kind from my earliest recollection (at just before one year) onwards to the present. I found myself as an experiencing being in Southern Australia (Victoria) in the early years of this century. I can ask: why there? And why then? These are the questions asked by Pascal. His answer was that it was God's will.

It is important to disclaim a solipsistic solution of the uniqueness of the self. Our direct experiences are, of course, subjective, being derived solely from our brain and self. The existences of other selves *are established* by intersubjective communication. The obvious explanation of my experienced uniqueness is that it results from my unique genetic endowment and that this is a necessary and sufficient condition. But this solution is untenable because of its unimaginable improbability (cf. Eccles, 1980). *If my uniqueness of self is tied to the genetic uniqueness that built my brain, the odds against myself existing are $10^{10\,000}$ against!* Hence I must reject this materialistic doctrine. What then determines the uniqueness of my psyche?

I have found that a frequent and superficially plausible answer to this question is the statement that the determining factor is the uniqueness of the accumulated experiences of a self throughout its lifetime. And this factor is also invoked to account for the distinctiveness of uniovular twins despite their genetic identity. It is readily agreed that my behaviour, my character, my memories and, in fact, the whole content of my inner conscious life are dependent on the accumulated experiences of my life; but no matter how extreme the change that can be produced by the exigencies of circumstance, I would still be the same self able to trace back my continuity in memory to my earliest remembrances at the age of 1 year or so, the same self in a quite other guise. Thus the accumulated experiences of a lifetime cannot be invoked as the determining or generating factor of the unique self, though naturally they will enormously modify all the qualities and features of that self.

Thus I am constrained to propose the creationist doctrine that each soul is a divine creation, that is attached *to the growing foetus at some time between conception and birth.* It is the existence of my inner core of unique individuality that *necessitates* the 'divine creation.' I submit that no other explanation is tenable, neither the genetic uniqueness with its impossible lottery nor the environmental differentiations, which do not *determine* one's uniqueness, but which merely modify it. The genetic make up actualized in the newly created life is random from an almost infinite number of possibilities. *The unique individuality comes from the infused soul.*

It is appropriate now to refer back to the analogy quoted above. The human brain (and body) is the computer and the conscious self, psyche, or soul is the programmer. The computer is built by genetic instructions and thus is a creation of biological evolution, being in World 1. The programmer is not in the biological world, but is in World 2 (cf. Figure 2). It is proposed that the 'programmer' is a divine creation.

Each of us is created as a unique psyche or programmer, and is given for life a unique computer, our brain, that is our sole means of receiving from and giving to the World in which we are immersed. This World comprises two major constituents: Worlds 1 and 3. There is a tendency in religious thinking to talk about persons or people, body plus soul, as the subjects of religious concern. However, I want to refer to the dualist-interactionist diagram of Figure 2 in *insisting* that World 2 (the soul or psyche) must *not* be blended in with the brain and body of World 1. There *is* an interface or frontier.

A very special problem is raised by identical twins that are completely distinct in their self-conscious experiences, alike as they appear to be to external observers. The same genetic endowments must therefore be compatible with different experiencing selves. *Evidently genetic constitution is not the necessary and sufficient condition for experienced uniqueness.* With the identical twins there would be a difference in the infused souls, just as with any other foetuses, non-identical twins for example.

Death and immortality?

On all materialist theories of the mind there can be no consciousness of any kind after brain death. Immortality is a non-problem. But with dualist-interactionism it can be recognized from the standard diagram (Figure 2) that death of the brain need not result in the destruction of the central component of World 2. All that can be inferred is that World 2 (the programmer) ceases to have any relationship with the brain (the computer) and hence will lack all sensory information and all motor expression. There is no question of a continued shadowy or ghost-like existence in some relationship with the material world, as is claimed in some spiritualist beliefs. What then can we say?

Belief in some life after death came very early to humankind, as is indicated by the ceremonial burial customs of Neanderthal man. However, in our earliest records of beliefs about life after death it was most unpleasant. This can be seen in the Epic of Gilgamesh or in the Homeric poems, or in the Hebrew belief about Sheol. Hick (1976) points out that the misery and unhappiness believed to attend the life hereafter very effectively disposes of the explanation that such beliefs arose from wish-fulfilment!

The idea of a more attractive after-life is a special feature of the Socratic dialogues, being derived from the Orphic mysteries. There was a particularly clear affirmation of immortality by Socrates in the *Phaedo* just before his death.

We normally have the body and brain to assure us of our identity but, with the departure of the psyche from the body and brain in death, none of these landmarks is available to it All of the detailed memory must be lost. If we refer again to Figure 2, memory is also shown located in World 2. I would suggest that this is a more general memory related to our self-identity, our emotional life, our personal life and to our ideals as enshrined in the values — in fact, the whole identity of the programmer. All of this should be sufficient for self-identity. Reference should be made to the discussion on the creation of the psyche by infusion into the developing embryo. *This divinely created psyche should be central to all considerations of immortality and of self-recognition.* With the disintegration of our computer and brain death, we have lost this wonderful instrument, the most intimate companion of a lifetime. Is there no further existence for the programmer?

The quest for meaning

Our life here on this earth and cosmos is beyond our understanding in respect of the Great Questions. We have to be open to some deep dramatic significance in this earthly life of ours that may be revealed after the transformation of death. We can ask: what does this life mean? We find ourselves here in this wonderfully rich and vivid conscious experience and it goes on through life; but is that the end? This self-conscious mind of ours has this mysterious relationship with the brain and, as a consequence, achieves experiences of human love and friendship, of the wonderful natural beauties, and of the intellectual excitement and joy given by appreciation and understanding of our cultural heritages. Is this present life all to finish in death, or can we have hope that there will be further meaning to be discovered?

Man has lost his way ideologically in this age. It is what has been called the predicament of humankind. I think that science has gone too far in breaking down man's belief in his spiritual greatness, as exemplified in the magnificent achievements in World 3, and has given him the belief that he is merely an insignificant animal that has arisen by chance and necessity in an insignificant planet lost in the great cosmic immensity. This is the message given to us by Jacques Monod in *Chance and Necessity*. I think the principal trouble with humankind today is that the intellectual leaders are too arrogant in their self-sufficiency. We must realize the great unknowns in the material make-up and operation of our brains, in the relationship of brain to mind, in our creative imagination and in the *uniqueness of the psyche*. When we think of these unknowns as well as the unknown of how we come to be in the first place, we should be much more humble. The unimagin-able future that could be ours would be the fulfilment of this our present life, and we should be prepared to accept its possibility as the greatest gift. In the acceptance of this wonderful gift of life and of death, we have to be prepared not for the inevitability of some other existence, but we can hope for the possibility of it.

The Heart of Healing

LARRY DOSSEY

Dr Larry Dossey M.D. is a physician of internal medicine and former Chief of Staff of Medical City Dallas Hospital in Dallas, Texas. He is the former co-chairman of the Panel on Mind/Body Interventions, Office of Alternative Medicine, National Institutes of Health, and is the current executive editor of the journal Alternative Therapies in Health and Medicine. He lectures widely in the United States and abroad. In 1988 he delivered the annual Mahatma Gandhi Memorial Lecture in New Delhi, India, the only physician ever invited to do so. He is the author of numerous articles and several books, including Space, Time & Medicine *(1982),* Beyond Illness *(1984);* Recovering the Soul *(1989);* Meaning & Medicine *(1991);* Healing Words: The Power Of Prayer & The Practice Of Medicine *(1993);* Prayer Is Good Medicine *(1996); and* Be Careful What You Pray For — You Just Might Get It *(1997). This lecture was delivered at the 1993 Mystics and Scientists conference.*

One of my heroes is Mahatma Gandhi. When the Mahatma was leading his country in its struggle for independence, a press conference was held in which a brash, young reporter asked, 'Mr Gandhi, what do you think about Western civilization?' Without hesitation Mr Gandhi responded: 'I think it would be an excellent idea!' If someone were to ask me, 'Dossey, what do you think about Western scientific medicine?' I would respond similarly: 'I think it would be an excellent idea.' Unfortunately we haven't had it yet; and the reason we have not yet had an authentically scientific medicine is that we have not had the integrity nor the courage to follow science wherever it leads. We have skimmed off the top, as it were, and we have focused on research findings that are harmonious with and supportive of our biases and our pre-existing world views, and we have left a much larger body of knowledge unattended. Thus Dr David Grimes was able to say in the *Journal of the American Medical Association* that 'much, if not most, of contemporary medical practice still lacks a scientific foundation.'[1]

This has had some disastrous consequences. Because we have neglected a large body of scientific evidence, we have largely amputated certain concepts from our notions of health and illness, among which are the effects of various manifestations of consciousness — thoughts, attitudes, emotions, perceived meanings, soul, spirit, heart. Never mind for the moment that these terms are imprecise and murky; they will declare their meanings to us as we go along. So I ask you to join me in a particular way of looking at this neglected information, and let us see together where it leads us. Our goal is to do the proper job of a scientist: to go *through* the information at hand, engaging it fully, letting it speak to us — not skirting it or avoiding it.

Mind and matter: The view from medical science

It is not difficult to understand why most physicians emerge from their training with no appreciation of anything 'higher' than the tissue and cells that make up the body. One can dissect cadavers endlessly and never find evidence for mind or consciousness, let alone soul or spirit. After four, eight, or twelve years of this sort of physicalistic orientation, it is no accident that physicians come to believe, as one famous neurophysiologist put it, 'The brain is where *all* the action is.' And that, as Bertrand Russell once put it, 'When I die I shall rot, and nothing of my ego shall remain.' Once can hardly find a better summary of this dismal point of view than that given by astronomer Carl Sagan in his book *The Dragons of Eden,* that the workings of the brain — what we sometimes call mind — are a consequence of its anatomy, physiology, and nothing more.[2] This point of view dominates Western bioscience. It is expressed in ways that sometimes seem almost humorous — as in a suggestion by Marvin Minsky, an expert in artificial intelligence at M.I.T., that the brain is nothing but a computer made of meat.

I frequently hear people say that this point of view does not affect them. They are on their 'spiritual path' and are relatively immune from these ideas. Yet I believe that most people are not so fortunate. Science is so imperious and monolithic — it really is the most powerful metaphor in our society — that its forlorn proclamations about the nature of consciousness gnaw at almost everyone to some degree. Deep within us, perhaps at an unconscious level, there is the suspicion, 'What if the scientists are correct? What if it *is* all over with the death of the brain?'

These messages are pervasive. Recently, *Time* magazine had a cover story for the Valentine's Day issue on the nature of love. The caption announced, 'Scientists are discovering that romance is a *biological* affair.' The heart was shown being poured from a test tube. The article noted the 'love chemicals' in the brain — serotonin, dopamine, norepinephrine, and phenylethylamine. The latter chemical is particularly interesting because its concentration is high in chocolate. That's presumably why we give bonbons to each other on Valentine's Day: it's 'all chemical.'

Accordingly, *all* human thought and feeling can presumably be reduced to chemical explanations. This has a rather chilling effect on the concept of self-help and individual responsibility, as exemplified in the so-called holistic or complementary health care movement. Feelings of self-empowerment are not substantial 'on their own,' for there really is no 'self' in 'self-help,' only the manifestations of our chemical underpinnings.

An implication of this point of view is that health and illness are entirely meaningless. They are only a function of what one's atoms are doing — and they are following the 'blind' laws of nature. These laws are inherently meaningless. Consequently, if we read any meaning into health and illness, we are suffering some sort of hallucination. 'Meaning' is a spurious concept; it comes from us, not nature.

Mind and matter: Another view

Although we continually hear arrogant, presumptuous statements from 'experts' that the relationship between brain, body, emotions, thoughts, and feelings has been resolved, that it's 'all brain,' one should not think that the debate on these matters is over. There is another side to this discussion which always goes unacknowledged. There have been stalwart, world-class, Nobel-calibre scientists, who have done science at the most fundamental levels, who have not agreed with this reductionistic, physicalistic view. Niels Bohr, for instance, whose name is virtually synonymous with modern physics, once said, 'We can admittedly find nothing in physics or chemistry that has even a remote bearing on consciousness.'[3] And his colleague, Werner Heisenberg, noted, 'There can be no doubt that consciousness does not occur in physics and chemistry, and I cannot see how it could possibly result from quantum mechanics.'[4] In other words, physics appears mute on the question of the origin of consciousness and its ultimate relationship to the material world. 'Consciousness' and 'physics' seem to occupy different domains; and one cannot conceivably 'work backwards,' beginning with a knowledge of the physical world, and 'disprove' consciousness. As the great Indian physicist D. S. Kothari stated in his seminal paper, 'Atom and Self,' consciousness is 'beyond physics.'[5]

Era I (Mechanical) Medicine

I ask you to drop back in time to the point in Western history when medicine first began to become scientific. This was the period of the 1860s, the decade of the American Civil War. Before this time, practically everything physicians did to help people was either worthless or actually harmful. Around this time physicians developed a profound case of physics envy, wanting to manifest in their profession the precision being demonstrated in classical, Newtonian physics. Thus began the first modern era of medicine, which we can designate as Era I or 'mechanical medicine.' We shall not dwell on this form of medicine. It still dominates, and should be familiar to anyone. I personally wish to distance myself from those who disparage this form of medicine and who want to dismantle and abandon it. To do so would, in my opinion, be foolhardy, inhumane, and downright silly. Our task is not to scuttle it but to find ways of transforming it — to learn to use it with caring, compassion, and love, qualities that have been notably lacking in this approach.

Era II (Mind-Body) Medicine

Beginning about fifty years ago, a new concept began slowly to organize. We came to acknowledge that the mind (if it existed!) might interact with the body in medically meaningful ways. But if it did so, the effects were generally considered negative, thus the designation 'psychosomatic *disease.*' Today this Era II or second era of scientific medicine is most commonly called 'mind-body' medicine.

Why does a second era claim our attention? Why can we not remain solidly in Era I? There is an avalanche of evidence demonstrating

	Era I	Era II	Era III
Space-time characteristic	Local	Local	Nonlocal
Synonym	Mechanical, material, or physical medicine	Mind-body or psychosomatic medicine	Nonlocal medicine
Description	Causal, deterministic, describable by classical concepts of space-time and matter-energy. Mind not a factor: 'mind' a result of brain mechanisms.	Mind a major factor in healing *within* the single person. Mind has causal power; is thus not fully explainable by classical concepts in physics. Includes but goes beyond Era I.	Mind a factor in healing both *within* and *between* persons. Mind not completely localized to points in space (brains or bodies) or time (present moment). Mind is unbounded in space and time and thus ultimately unitary or one. Not describable by classical concepts of space-time or matter-energy.
Examples	Any form of therapy focusing solely on effects of *things* on the body are Era I approaches. Almost all forms of modern medicine – drugs, surgery, irradiation, CPR, etc. – are included.	Any therapy emphasizing the effects of consciousness *solely* within the individual body is an Era II approach. Psychoneuro-immunology, counselling, hypnosis, biofeedback, relaxation therapies, and most types of imagery-based 'alternative' therapies are included.	Any therapy in which effects of consciousness bridge between different persons is an Era III approach. All forms of distant healing, intercessory prayer, 'psychic' and shamanic healing, diagnosis at a distance, and transpersonal imagery are included.

Table 1. Medical eras.

profound instances of the interaction of mind and body that are relevant to human health. Examples come from all cultures. Consider, for instance, certain religious rites such as those engaged in by a certain Islamic Sufi sect. After a night of fasting, meditation, and prayer, they are in an altered state of consciousness. In such a state, they frequently inflict punishments on themselves as a way of honouring God. This takes many forms, such as skewering their tongue and bodily organs such as the liver with metal objects; hammering sharp objects into solid bone; chewing up a Coke bottle; and holding their face in flames. They do not bleed, do not become infected, so not experience obvious pain, and heal amazingly rapidly, often without scarring. These events, which are being investigated by a group at the University of Durham in England, defy the known laws of physiology, yet are extremely common within this particular religious sect.

Astonishing mind-body events occur not only in exotic cultures but in our own as well. In fact, they are so common we cease to notice them. Let me illustrate how pervasive they are by referring to the commonest cause of death in our culture, coronary artery disease. This illness kills more people annually than all the other causes of death combined. We develop this problem, we say, by having one or more of the major risk factors present — diabetes mellitus, high blood cholesterol, cigarette smoking, or high blood pressure. It is true that the presence of these risk factors increases the risk of having heart disease; if they are present, thus, we should consider eliminating them. Yet epidemiological work dating to the early seventies documents that the majority of persons under the age of fifty who have their first heart attack have *no* major risk factors present.[6] (This observation must currently be tempered. We have recently defined 'normal' cholesterol and blood pressure downward, so that many more people today are 'abnormal' than in the early seventies when this observation was made. Still, it is likely that the predictive value of the major risk factors is oversold in the mind of the general population. It is *not* true that one will necessarily escape a heart attack if they are eliminated.)

As further evidence of the 'strangeness' surrounding heart disease, Dr James Muller at the National Institute of Health discovered that more heart attacks occur between eight and nine AM than at any other time.[7] And subsequent work has demonstrated that more heart attacks occur on *Monday* than on any other day — the so-called 'Black

Monday syndrome.' This is one of the most bizarre findings not just in medicine but in the entire world of biology. As far as we know, *human beings are the only species on the face of the earth who manage to die more frequently on a particular day of the week.*

What is unique about Monday morning, eight to nine AM? I think it would not be irrational to suppose that it just *might* have something to do with the beginning of the work week. In support of this, a state-wide survey done in the early seventies in the state of Massachusetts showed that the best predictor of a heart attack was *none* of the major risk factors but the level in a person's life of *job dissatisfaction.*

Therefore we are justified in asking: What does Monday morning, 8-9 AM, *mean* to someone? What does this moment in time symbolize, represent, or stand for? If the meaning is negative, we should consider this to be a major risk factor for the commonest cause of death in our society.

Researchers in this field, such as Dr Robert Karasek and his colleagues, have proposed the existence of the 'Sisyphus reaction' to account for these job-related events.[8] Sisyphus was the Greek mythological figure who was condemned forever to push the boulder up the hill, only to have it roll down again, requiring that the action be endlessly repeated. This 'syndrome of joyless striving' is found in many modern occupations. Researchers have identified two major components which, if present, increase one's risk of sudden death on the job from heart attack: *psychological stress* and *lack of control.* Several occupations have been identified in which there exists an increase in the risk of sudden death on the job. Among them are waiters in very busy restaurants, certain data processors and assembly line workers, and gas station attendants at busy gas stations. It should be noted that psychological stress, of itself, is not enough to put one at risk; many people thrive on stress and never become ill. Lack of control of the job load is the other essential ingredient. When these two factors are combined, the result is a joyless, meaningless, frustrating, helpless, and powerless feeling that, if sustained, can be toxic and fatal to the human heart.

Meaning therapy
If negative meanings can harm our health, this raises the possibility that if we somehow convert them from negative to positive, they might improve our health and might even save our life. Thus I wish to

describe a new kind of therapy — 'meaning therapy' — for which much clinical data can be put forward.

One of this century's landmark clinical studies was an example of meaning therapy in action. It was done by Dr Dean Ornish in California and was published in England's prestigious *Lancet*. Ornish treated a group of men who had severe coronary artery disease. Many of them had already had coronary artery bypass operations. Many had angina at rest — chest pain doing nothing — which is a terrible prognostic sign. Ornish placed them on a program consisting of three parts — a very low-fat diet, a program of yoga and walking, and group therapy. The patients and their wives came together once a week for a year, and essentially did 'meaning therapy.' They discussed the *meaning* of heart disease — what it *means* to realize they've already had maximal therapy including surgery; that they will never return to work; that if they make love with their wives they may not survive; that they're only waiting around for 'the big one.' Ornish discovered that the chest pain disappeared in a matter of days to weeks. After one year he re-studied these men with sophisticated catheterization techniques and proved that something had occurred that has been considered impossible in modern medicine: the heart disease had begun to reverse. The concrete-like obstructions in the coronary arteries had actually begun to shrink and go away. This was not a temporary improvement; four years later, he has shown that this process continues to occur.

This high-touch, low-tech, low-budget approach to our commonest killer is currently under close examination as an alternative to coronary artery bypass surgery. And for excellent reasons. In 1992, 300,000 such operations were performed in the United States, costing some $8 billion. None of these operations did anything to change the underlying disease process, which continues to progress. Ornish's program of meaning therapy, diet, and exercise actually *reverses* the disease process.

The positive effects of meaning therapy are not confined to heart disease. Dr David Spiegel, a psychiatrist at Stanford Medical School, did a controlled study in which he brought together a group of women who had metastatic breast cancer to engage in essentially the same process described above, that of examining the meanings they perceived surrounding their illness.[9] The women had already undergone conventional therapy including surgery, chemotherapy, and irradiation. (Spiegel, a skeptic of the role of the mind in the cancer process, designed this study to lay to rest the 'counterculture' idea of the mind's role in cancer.) These women met once weekly for one year only. After ten years, Spiegel examined the data and found that the meaning therapy group on average survived *twice* as long following diagnosis as did women treated only conventionally. Moreover, there were three ten-year survivors in the entire study who presumably were completely cured; all three belonged to the meaning therapy group.

It is almost certain that if it were a new drug or surgical procedure that were being evaluated in the Ornish and Spiegel studies, the new treatment would have been heralded as a 'medical miracle' or a 'breakthrough.'

A study was recently reported from Yale Medical School in which researchers asked the question, 'What is the best way to predict who is going to be alive at the end of the next decade?'[10] Physicians would ordinarily want to assess certain time-honoured factors in order to answer this question accurately, such as one's past medical history, one's family history, and the results of one's physical examination and laboratory tests as one enters the upcoming decade. The researchers discovered, however, that the best predictor of longevity over the next ten years was *none* of these factors, but the answer people gave to a simple question: 'What do you *think* about your health?' If people gave a strongly negative answer, they had a seven-fold increase in the risk of dying over the next decade, compared to someone who gave a positive answer, all other things being equal. This study correlated with the results of four previous studies, encompassing tens of thousands of people. This is powerful evidence that *we live out our meanings,* that perceived meanings make the difference in life and death.

Era III (Non-local) Medicine[11]

Many believers in holistic, alternative, or complementary medicine think that Era II or 'mind-body' medicine is as exotic or 'far out' as medicine can get. After all, what could be more dramatic than using consciousness to reverse coronary artery disease and double survival rates in certain cancers? Yet there is substantial reason to postulate another era in the scientific march of medicine — Era III, or 'non-local' medicine.

'Non-local,' as I wish to use the term, refers to the relationship between the mind, the brain, and the temporal sequence. First, a *local* concept of these relationships should assert that the mind is localized or confined to the brain, the brain is localized to the body, and the mind, brain, and body are localized to the present moment. This picture is virtually unquestioned in science and is accepted by most laypersons. It is common sense.

In contrast, a *non-local* view would contend that the mind may not be localized or confined to the brain, nor perhaps to the present moment. The mind, in other words, would be unbounded in space and time. Now, if one supposes that mind-body interaction is real — and that's what Era II medicine is all about — and that minds are non-local, we encounter the possibility that your mind may be able to affect my body, and my mind may be able to affect yours.

Outrageous? Perhaps; but empirical evidence suggests that such events are not only possible but commonplace.

Transpersonal imagery. At the Mind Science Foundation in San Antonio, Texas, Dr William G. Braud, Dr Marilyn Schlitz, and their colleagues have examined the impact of the mental imagery of one person on the physiology of a distant person, who is unaware that such imagery is being directed to them. These studies follow the criteria of excellent science. Hundreds of trials have been done.

Results indicate, in brief, that the mental images of one person seem capable of 'reaching out' across space and causing 'robust' changes in a distant individual's physiological processes that are comparable to the effects of one's own mental images on one's body.[12]

Distant EEG correlations. Several experiments have been done in which two distant people are 'wired up' for electroencephalograms or EEG recordings.[13] In the baseline state, there is no correlation between the patterns. Then the experimenters tell the subjects to attempt to come together emotionally, to develop empathy with each other, even though they are separated. When they indicate they have done so, the EEGs begin frequently to cycle together, often appearing identical. In a variation of the experiment, one of the subjects has even been shielded in a Faraday-type cage, which eliminates, for practical purposes, all electromagnetic irradiation.

Could this be indirect evidence of the claims of 'psychic healers' throughout history of the power of love, empathy, and compassion to 'reach out' and bring about healthful change in a distant person?

I fully realize the heretical nature of these observations. According to mainstream science, these events can't happen and therefore they don't happen. Any evidence to the contrary must be due to faulty observation, naïveté, or downright fraud. In fact, for three-hundred years one of the most unkind things one scientist could say about another is that 'he believes in action at a distance.' When the young Kepler proposed that the Earth's tides were caused by lunar gravity, Galileo ridiculed him. He called Kepler's ideas 'the ravings of a madman' because they invoked 'action at a distance.' Although Kepler was vindicated, the pejorative insult has not died but is still commonly used.

Prayer. There is a large number of controlled laboratory experiments showing that intercessory prayer has a significant effect in a host of biological organisms. Daniel J. Benor, an American psychiatrist, has extensively reviewed perhaps all the studies in prayer-type healing published in the English language. He found a total of 131 studies, of which 56 demonstrated a p value of $< .01$, and an additional 21 had a p value between .02 and .05. This evidence is available in Benor's *Healing Research.*[14] This evidence and its implications for medicine, health, and healing have also been reviewed in my books *Healing Words: The Power of Prayer and the Practice of Medicine,* and *Prayer Is Good Medicine.*[15] I regard this information collectively as one of the best-kept secrets in medical science. If taken seriously, this information could revolutionize our understanding of the nature of consciousness, the relationship between mind and brain, and the actual dynamics of healing. This evidence is too vast to be discussed here. It is, however, creating quite a stir in the United States. As of mid-1997, eleven premier medical schools in the United States have elective courses devoted to the role of spirituality and clinical practice, and half have expressed interest in developing such courses.[16]

Distant diagnosis. There is a sizeable body of evidence suggesting that people can make diagnoses at great distances. In the United States this work is featured most dramatically in the book, *The Creation of Health,* by Dr Norman Shealy and Carolyn Myss. Dr Shealy is the Harvard-trained neurosurgeon who founded the American Holistic Medical Association. He began to work with Carolyn Myss a few years ago. In brief, Dr Shealy would have a patient in his office in Missouri, and would phone Myss in New Hampshire (about half way across the United States). He would provide her with the first name and the birth date of the patient, and she would provide him with the diagnosis. In the first hundred cases she is 93 percent correct. This is rather astonishing; I know of no internists who are 93 percent correct with such a paucity of information. Could Shealy have 'telepathically transmitted' the correct diagnosis to Myss? We cannot say with certainty. Even if he did so, this is hardly less remarkable![17]

Telesomatic events. 'Telesomatic' comes from Greek words meaning 'distant body.' Hundreds of cases have been described in which distant people share symptoms and sometimes actual physical changes, without knowing what is happening at the time to the other distant person.

As an example, John Ruskin reported in 1899 a telesomatic event involving Arthur Severn, a landscape painter. Unable to sleep, Mr Severn arose early one morning to go to the lake for a sail. Mrs Severn, who stayed in bed, was suddenly awakened with the sensation of a severe blow to the mouth. Later Mr Severn returned, with a bloody handkerchief held to his mouth. The wind had suddenly increased, forcing the tiller around, hitting him in the mouth and almost knocking him out of the boat.

Skeptics will see nothing unusual in this — just one of those 'funny coincidences.' Certainly these cases are not 'science' but 'stories.' (There are two ways in which physicians respond to 'stories.' If one looks pejoratively on the story it is called an anecdote; if positively, it becomes a case history.) It is true that these events cannot be compelled to happen in the laboratory so that we can conveniently study them. But in spite of their unpredictability, there are two features of telesomatic events that compel our attention. First, they are extraordinarily common. Second, they display an internal consistency that is simply stunning. At some point, they begin to sound alike. They involve people who are always at a distance and who are *empathic* with each other — the felt quality we encountered above in the distant EEG correlation studies wherein physiological traits of widely separated persons began to attune with each other. Telesomatic events occur most classically between parents and children — the mother who feels a suffocating feeling and 'just knows' her child is drowning, and rushes home just in time to drag the child from the swimming pool and save its life. They also occur commonly between spouses, siblings (particularly twins), and lovers.[18]

Remote viewing. Studies in remote viewing have been replicated in several laboratories, most dramatically in the past decade at Princeton University's PEAR, the Princeton Engineering Anomalies Research laboratory, under the direction of Princeton's former dean of engineering, Robert G. Jahn. In these studies a 'sender' remains on-site at Princeton and a 'receiver' journeys to a distance, sometimes to the other side of the Earth. A computer at Princeton then randomly selects from a large data bank an image or picture, which is presented to the sender, who registers it in the mind and tries mentally to 'send' its contents to the receiver. The receiver then draws 'what comes through.' The drawing is eventually fed into the computer, which decides whether there is a hit or a miss between what was sent and what was received.

Many times the image gets through in stunning, camera-like detail. Amazingly, in the majority of instances the receiver 'gets' the information up to three days *before* it is sent, *before* it is even selected by the computer.[19].

These findings have ignited a furore within science, and skeptics and cynics have expended enormous amounts of energy attempting to discredit this information. It is not difficult to see why these studies have generated such intellectual indigestion. They invoke those perennial fighting words: mental action at a distance.

Mind: Toward a non-local model

In sum, the non-local, Era III-type events we've examined reveal two qualities of the mind that demand our attention. Studies in transpersonal imagery, distant EEG correlations, distant diagnosis, telesomatic events, and remote sensing suggest that some aspect of the psyche is *non-local* — that is, it cannot be confined to specific points in space such as brains or bodies, nor to specific points in time such as the present moment.

If some aspect of the mind is genuinely non-local, the implications are profound. Non-locality in space and time does not mean 'quite large' or 'a very long time.' Non-locality implies *infinitude* in space and time, because *a limited non-locality is a contradiction in terms.*

In the West we have traditionally defined 'soul' as something unborn and which does not die, something infinite in space and time and therefore omnipresent, eternal, immortal. *That is why making a non-local model of the mind is essentially an act of recovering the soul.* Never mind for the moment that some spiritual traditions such as Buddhism are said not to believe in souls. Buddhists nonetheless believe in one's eternal 'Buddha nature,' which bears the strongest resemblance to the very broad sense in which we are describing 'soul' — not as an entity or thing but as a non-local quality of mind.

It may be no exaggeration to say, therefore, that for the very first time in human history we have indirect, empirical evidence for the existence of 'soul.' While it is true that we have no 'soul meters' we can plug into individuals to get a direct readout on the presence or absence of 'soul,' we have the next best thing: manifestations of the mind which display themselves non-locally in controlled laboratory settings. We have never before had such evidence. Belief in the soul has always been judged a matter of sheer faith. Today it is no longer *just* a matter of faith.

Perhaps these developments are a hopeful sign that the longstanding divide between science and spirituality may diminish, and that scientific and religious thought may learn to stand side by side, neither trying to demolish or usurp the other.

Emerging theories

'Non-local mind' is a term I introduced in 1989 in my book *Recovering the Soul.*[20] According to this concept, as we've noted, consciousness cannot be completely localized or confined to specific points in space, such as brains or bodies, or to discrete points in time, such as the present moment. A mind that is non-local might work *through* the brain and body, without being limited to them.

This picture of the mind is gaining increasing attention within the scientific community. Many scientists do not believe that the type of events we've examined are in conflict with current scientific theory. For example, the eminent physicist Gerald Feinberg, speaking of non-local distant mental events, says, '... [If] such phenomena indeed occur, no change in the fundamental equations of physics would be needed to describe them.'[21]

The model of consciousness that is needed to accommodate distant mental intentions, whether positive or negative, is one that recognizes a non-local quality of the mind. Several models of consciousness have recently been proposed by eminent scientists that embody this quality of the mind. For example:

- Physicist Amit Goswami of the University of Oregon's Institute of Theoretical Science has proposed his Science Within Consciousness (SWC) theory, in which consciousness is recognized as a fundamental, causal factor in the universe, not confined to the brain, body, or the present.[22]

- David J. Chalmers, a mathematician and cognitive scientist from University of California, Santa Cruz, also has suggested that consciousness is fundamental in the universe. It is not derived from anything else, and cannot be reduced to anything more basic. Consciousness, Chalmers suggests, is perhaps on a par with matter and energy.[23] His view frees consciousness from its local confinement to the brain, and opens the door for non-local, consciousness-mediated events such as we've discussed.

- Physicist Nick Herbert has long proposed a similar view. He suggests that consciousness abounds in the universe and that we have seriously underestimated the 'amount' of it, just as early physicists drastically underestimated the size of the universe.[24]

- Nobel physicist Brian D. Josephson, of Cambridge University's Cavendish Laboratory, has proposed that consciousness makes possible 'the biological utilization of quantum non-locality.' He believes that non-local events exist not only at the subatomic level, but, through the actions of the mind, can be amplified and emerge in our everyday experience as distant mental events of a broad variety.[25]

- Rupert Sheldrake, the British botanist, has proposed a non-local picture of consciousness in his widely known 'hypothesis of formative causation.' Sheldrake sees great promise in his model for distant, mental events such as intercessory prayer and for negative mental intentions as well.[26]

- Systems theorist Ervin Laszlo has proposed that non-local, consciousness-mediated events such as intercessory prayer, telepathy, precognition, and clairvoyance may be explainable through developments in physics concerning the quantum vacuum and zero-point field.[27]

- The late physicist David Bohm proposed that consciousness is present to some degree in everything. 'Everything material is also mental and everything mental is also material,' he states. 'The separation of the two — matter and spirit — is an abstraction. The ground is also one.'[28] Bohm's views, like the above hypotheses, liberate consciousness from its confinement to the body and make possible, in principle, the distant, non-local phenomena we've examined.

- Robert G. Jahn, former dean of engineering at Princeton University, and his colleagues at the Princeton Engineering Anomalies Research lab, have proposed a model of the mind in which consciousness acts freely through space and time to create actual change in the physical world. Their hypothesis is based on their experimental evidence, which is the largest database ever assembled of the effects of distant intentionality.[29]
- Mathematician C.J.S. Clarke, of the University of Southampton's Faculty of Mathematical Studies, has proposed that 'it is necessary to place mind first as the key aspect of the universe.' Clarke's hypothesis is based in a quantum logic approach to physics, and takes non-locality as its starting point.[30]

Although these views are recent, they are part of a long tradition within modern science. Many of the greatest scientists of this century have been cordial to an extended, unitary model of the mind, which is sort of picture that permits the sort of distant mental intentions we've been examining.[31] This shows that a non-local view of consciousness is not a fringe or radical idea, as critics often claim

For example, Nobel physicist Erwin Schrödinger entertained this idea in Berlin in the twenties and thirties. He said, 'Mind by its very nature is a *singulare tantum;* I should say, the overall number of minds is just one.'[32] Why 'just one'? If minds are genuinely non-local, their boundaries are not absolute. Therefore there cannot be six billion separate minds on the Earth today; in some sense and at some level there can only be a single mind.

Sir Arthur Eddington, the eminent astronomer-physicist: 'The idea of a universal Mind or Logos would be, I think, a fairly plausible inference from the present state of scientific theory; at least it is in harmony with it.'[33]

Sir James Jeans, the British mathematician, astronomer, and physicist: 'When we view ourselves in space and time, our consciousnesses are obviously the separate individuals of a particle-picture, but when we pass beyond space and time, they may perhaps form ingredients of a single continuous stream of life. As it is with light and electricity, so it may be with life; the phenomena may be individuals carrying on separate existences in space and time, while in the deeper reality beyond space and time we may all be members of one body.'[34]

Neither is this idea foreign to poets. William Butler Yeats echoed Schrödinger: 'The borders of our minds are ever shifting, and many minds can flow into one another and create or reveal a single mind, a single energy.'

In summary, non-local mind — mind beyond body, mind beyond time — is supported today by a four-letter word that makes all the difference: *data*. In digesting this information, we might well be guided by the view of philosopher John Searle: 'At our present state of the investigation of consciousness we *don't know* how it works and we need to try all kinds of different ideas.'[35]

Implications for medicine

What are the immediate, practical, medical consequences of such considerations? Almost all physicians are trained to hold back the clock, to extend life. A long life is better, we say, than a short one. This reflects our belief that each person is quintessentially a *local* creature, confined to his or her brain/body in space, moving locally along the river of time. At some tragic point in the future our existence will forever end — the typical, local view. Patients participate in this view. Not only do they want their physicians to struggle valiantly to *extend* their time, they try mightily to *hold on to* time. This creates an insoluble problem, which could be expressed thus: 'After the miracles, what then?' No matter how powerful modern medicine becomes, sooner or later the miracles will run out. That is why the beginning assumption of medicine is *tragedy:* we know ahead of time that medicine will fail.

Eternity medicine
In contrast, if we take seriously a non-local conception of the human mind, we can conceive of what could be called *eternity medicine*. This stands in stark contrast with the *temporal medicine* with which we are currently besotted. Eternity medicine transcends the finality of death, in contrast to temporal medicine. It is not lodged in tragedy; it rests on the realization that the most essential part of who we are is *infinite* in space and time; is thus eternal and immortal; is unborn; is incapable of death.

The Divine within
This recognition would also make possible a different conception of our relationship to the Absolute (God, Goddess, Allah, Brahman, the

Tao, the Universe, Cosmos, and so on). We define the Absolute as omnipresent, infinite in space and time, eternal, immortal — the very qualities manifested in our own non-local nature. Thus we share qualities with the Absolute — the 'Divine within' concept exemplified in many great wisdom traditions. The Hindu phrase, 'Tat tvam asi!' or 'Thou art that!' is a typical example of this realization.

It is unfair to characterize Western science as uniquely stubborn in its resistance to these ideas. The record of Western religions is no better. Indeed, the idea of the 'Divine within' has perhaps been the most difficult concept for the Western religions to assimilate. Those who have espoused it — Germany's great mystic of the thirteenth century, Meister Eckhart, is only one example — have all too frequently been branded as heretics and blaphemers and not infrequently killed. Even today Western religious traditions continue to resist these considerations. Their emphasis remains on the unworthiness and the sheer locality of humans, not on the glorious, divine qualities implied by a non-local view of humankind.

A complementary approach

Just because we embrace 'eternity medicine,' we do not have to abandon the Era I, mechanical approaches currently in vogue. This is vastly misunderstood. Many people believe that the recognition of their intrinsic, non-local nature constitutes a mandate to invade all the hospitals, unplug all the ventilators, and unhook all the IVs. Yet it is nowhere written down that we must pursue only one approach. We can still opt for mechanical, Era I-type approaches. But if we choose to use them following an Era III awakening, we now use them with a difference — with a twinkle in our eye and our tongue in our cheek, knowing that, should the Era I approaches fail, as they eventually will, *there is no tragedy* because the most essential part of us in principle cannot die. This might allow for yet another miracle — an infusion of a much-needed lightness of heart and perhaps humour in the modern medical effort.

Spiritual understanding and physical health

When we penetrate deeply into a realization of our non-local nature, there is a tendency to believe that we shall leave disease and illness behind; that physical health is some sort of guarantee once we have

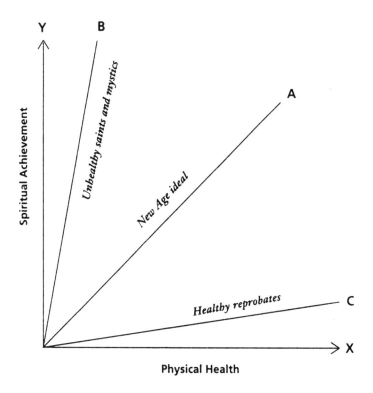

Figure 1. The level of spiritual achievement is plotted along the y-axis, and the degree of physical health is plotted along the x-axis. The New Age ideal — the notion that spiritual achievement and physical health are always correlated — is illustrated by line A. This notion presupposes that for every gain in spirituality there is a corresponding gain in physical health, without exception — a one-to-one, straight line, linear, invariant relationship. That this is not the case is illustrated by lines B and C. Line B shows that saints and mystics may be high spiritual achievers but may have poor physical health. Line C illustrates the opposite: people with little spiritual sensitivity — so-called spiritual reprobates — can enjoy extremely good physical health.

done our 'spiritual homework.' The facts say otherwise. Even the most saintly, God-realized, spiritually evolved persons die frequently of the pesky problems that plague us lesser mortals. Three of the holiest people of the twentieth century have died of cancer — Krishnamurti, cancer of the pancreas; Ramana Maharshi, cancer of the stomach; Suzuki Roshi, cancer of the liver. Bernadette, who saw the vision of the Virgin at Lourdes, died of bone cancer or disseminated tuberculosis at age 33. Jesus Christ died at age 33 of trauma. The Buddha died of food poisoning, having been fed tainted meat at what proved to be his last meal. These deaths, many of which were painful and grotesque, contain a lesson: one can be highly evolved spiritually and become very ill.[36]

In the ninth chapter of the Gospel of John, the disciples are following Jesus as they encounter the man blind since birth. The disciples ask, Who hath sinned, this man or his parents, that he should be born blind? This question is surprisingly contemporary. It reflects the idea that spiritual shortcomings and physical illness go hand-in-hand, the notion that ill health is direct evidence of spiritual failure. But Jesus did not agree. Before performing a healing, he stated that no one had sinned, neither the congenitally blind man nor his parents, but that the works of God should be made manifest through this man. This is an example of the complete *dissociation* between spiritual imperfection and illness, a lesson also exemplified in the health histories of the saints and mystics above.

It is important to point out these relationships, illustrated in Figure 1, because we seem to be suffering culturally from an epidemic of New Age guilt — the sense of failure, shame, and self-blame that can occur when one becomes ill while involved in authentic spiritual growth. Might our human capacities change in the future? Perhaps, at some future stage of human evolution, we may be able completely to manage our healing with 'the powers of consciousness' and dwell only in Eras II and III. Currently, however, this is not possible for most of us — not even for the highly evolved spiritual geniuses, as we saw. If we recognize these facts, we might spare ourselves the unnecessary sense of guilt and failure when we become ill in spite of considerable spiritual understanding.

Let us not, therefore, draw simplistic formulas in which we equate spiritual understanding and physical health; that would be an abuse of the information we've examined. Let us focus instead on the central realizations that shine forth from the non-local perspective:

1. that our most essential qualities are non-local and therefore immortal and eternal;
2. that illness and death are problematic only from a local perspective;
3. that tragedy belongs only to temporality and locality, not to eternity;
4. that 'the Divine within' is not poetic metaphor but an accompaniment of our non-local nature.

The universe is friendly!

Einstein once remarked that the most important question one can ask is, 'Is the Universe friendly?' Many responses have been given to this question. We noted above Bertrand Russell's response: 'When I die I shall rot, and nothing of my ego shall remain.' And there is the response from Buddhism: 'If you die before you die, then when you die you will not die.' I believe these two responses are compatible. Lord Russell was surely correct: when we die, nothing of our ego, our small self, *will* remain. But that is not the whole story, as the Buddhist aphorism suggests. For if we are willing to pass from the ego-oriented, *local* way of defining ourselves to the *non-local* mode of awareness, we know there is no death. This change in perspective justifies a further observation of Einstein's: 'The beauty of it is that we have to content ourselves with the recognition of the *miracle,* beyond which there is no legitimate way out.'

References

1. David A. Grimes, 'Technology Follies: The Uncritical Acceptance of Medical Innovation,' *Journal of the American Medical Association,* 269:23, June 16, 1993, pp. 3030-33
2. Carl Sagan, *The Dragons of Eden* (New York: Random House, 1977), p.7.
3. Niels Bohr, quoted in Werner Heisenberg, *Physics and Beyond,* A.J. Pomerans, trans. (New York: Harper and Row, 1971), pp. 114-115.
4. Werner Heisenberg, *Physics and Beyond,* A.J. Pomerans, trans. (New York: Harper and Row, 1971), p.114.
5. D.S. Kothari, 'Atom and Self.' The Mehnad Saha Medal Lecture 1978. *Proceedings of the Indian National Science Academy,* Part A, Physical Science 46:1 (1980), pp.1-28.
6. C.D. Jenkins, 'Psychological and Social Precursors of Coronary Artery Disease,' *New England Journal of Medicine,* 284 (1971), pp. 244-55.
7. Gina Kolata, 'Heart Attacks at 9:00 am,' *Science* 233 (July 25, 1986), pp. 417-18.
8. R.L. Karasek, T. Theorell, J. E. Schwartz, P. L.Schnall, C. F. Pieper, J. L. Michela. 'Job Characteristics in Relation to the Prevalence of Myocardial Infarction,' *American Journal of Public Health,* 1988; 78(8): pp. 910-16.
9. D. Spiegel, J.R. Bloom, H.C. Kraemer, and E. Gottheil. 'Effect of Psychosocial Treatment on Survival of Patients with Metastatic Breast Cancer,' *Lancet.* 1989; 2(8668): pp.888-91.
10. E.L. Idler and S. Kasl, 'Health Perceptions and Survival: Do Global Evaluations of Health Status Really Predict Mortality?' *Journal of Gerontology,* 1991; 46: S55-S65.

11. Larry Dossey, *Meaning and Medicine* (New York: Bantam, 1991), pp.178-93.
12. William Braud and Marilyn Schlitz, 'A Methodology for the Objective Study of Transpersonal Imagery,' Journal of Scientific Exploration, 1989; 3(1); pp.43-63. Also: William Braud and Marilyn Schlitz, 'Consciousness Interactions with Remote Biological Systems: Anomalous Intentionality Effects,' *Subtle Energies,* 1991; 2(1): pp.1-46.
13. J. Grinberg-Zylberbaum, J. M. Delaflor, L. Attie, A. Goswami. 'The Einstein-Podolsky-Rosen Paradox in the Brain: The Transferred Potential,' *Physics Essays,* 1994; 7(4): pp.422-28.
14. Daniel J. Benor, *Healing Research,* vol. 1 (Munich: Helix Verlag, 1993).
15. Larry Dossey, *Healing Words* (HarperSanFrancisco, 1993); *Prayer Is Good Medicine* (HarperSanFrancisco, 1996).
16. Information about these developments may be obtained from the National Institute for Healthcare Research, 6110 Executive Blvd., Suite 908, Rockville, Maryland, USA 20852.
17. C. Norman Shealy and Caroline M. Myss, *The Creation of Health* (Walpole, NH: Stillpoint, 1988).
18. Ian Stevenson, *Telepathic Impressions: A Review of 35 New Cases* (Charlottesville: University Press of Virginia, 1970).
19. Robert G. Jahn and Brenda J. Dunne, *Margins of Reality* (New York: Harcourt Brace Jovanovich, 1987).
20. Larry Dossey, *Recovering the Soul* (New York: Bantam, 1989).
21. Gerald Feinberg, 'Precognition — A Memory of Things Future.' In: L. Oteri (ed.), *Quantum Physics and Parapsychology,* (New York: Parapsychology Foundation, 1975), pp.54-73.
22. Amit Goswami, *The Self-Aware Universe: How Consciousness creates the Material World* (New York: Tarcher/Putnam, 1993). See also, 'Science Within Consciousness: A Progress Report.' Talk delivered at a seminar on consciousness, University of Lisbon, Portugal, 1996.
23. David J. Chalmers, 'The Puzzle of Conscious Experience,' *Scientific American* 1995; 273(6): pp.80-6. See also *The Conscious Mind* (New York: Oxford University Press, 1996).
24. Nick Herbert, *Quantum Reality* (New York: Dutton, 1986). Also *Elemental Mind* (New York: Dutton, 1993).
25. Brian D. Josephson and F. Pallikara-Viras, 'Biological Utilization of Quantum Nonlocality,' *Foundations of Physics,* 1991; 21: pp.197-207.
26. Rupert Sheldrake, *A New Science of Life.* (Los Angeles: Tarcher, 1981); *The Presence of the Past* (New York: Times Books, 1988).
27. Ervin Laszlo, *The Interconnected Universe: Conceptual Foundations of Transdisciplinary Unified Theory* (River Edge, NJ: World Scientific Publishing Co., 1995).
28. David Bohm quoted in Renée Weber, *Dialogues with Scientists and Sages: The Search for Unity* (London: Arkana, 1990), pp.101, 151.
29. Robert G. Jahn and Brenda J. Dunne, *Margins of Reality: The Role of Consciousness in the Physical World* (New York: Harcourt Brace Jovanovich, 1987).
30. C.J.S. Clarke, 'The Nonlocality of Mind,' *Journal of Consciousness Studies.* 1995; 2(3): pp.231-40.

31. Ken Wilber, ed. *Quantum Questions: The Mystical Writings of the World's Great Physicists* (Boston: Shambhala, 1984).
32. Erwin Schrödinger, *What Is Life? and Mind and Matter* (London: Cambridge University Press, 1969), p.145.
33. Arthur Eddington, 'Defense of Mysticism,' in *Quantum Questions: The Mystical Writings of the World's Great Physicists,* Ken Wilber, ed. (Boston: Shambhala, 1984), p.206.
34. James Jeans, *Physics and Philosophy* (New York: Dover, 1981), p.204.
35. John Searle, cover quotation, *Journal of Consciousness Studies.* 1995; 2(1).
36. Larry Dossey, 'Saints and Sinners, Health and Illness,' *Healing Words* (HarperSanFrancisco, 1993), pp.13-46.

Mysticism and Spirituality

Gateway to the Infinite

SIR GEORGE TREVELYAN

Sir George Trevelyan grew up against a background of liberal politics and progressive thought. He read History at Cambridge, and taught at Gordonstoun. On retirement from the army he became principal of Attingham Park, the Shropshire Adult College where he did his pioneering work in the teaching of spiritual knowledge as adult education. On his retirement in 1971 he founded the Wrekin Trust. Books include A Vision of the Aquarian Age, Operation Redemption, Magic Casements, Summons to a High Crusade *and* Exploration into God. *This lecture was delivered at the 1978 Mystics and Scientists conference.*

Our conference is something of a phenomenon. Firstly it has drawn a notable and distinguished response, but, more important, it indicates a trend of change in our thinking, a remarriage of what was once united and has for centuries been divorced, This separation of mysticism and scientific thinking was itself a symptom of evolving consciousness. In these opening words I do honour to Goethe, that great scientist/poet who knew that in our approach to nature, the real wisdom can only be achieved if the imaginative vision and intuition of the artist is united with the acute observation and analysis of scientific method. He saw that intellect alone cannot really apprehend life; but that, in order to enter into direct and immediate experience of the life and being within the form, it was necessary to develop new, or latent, faculties of perception.

His discovery of the metamorphosis of plants, for instance, was achieved through his developing an exact imaginative observation that enabled him to move in thought through the complete cycle of plant growth, from seed through leaf, flower and fruit and back to seed.

When we achieve this, the plant begins to reveal its secrets. It 'speaks' to our thinking.

It was Rudolf Steiner who carried this technique a notable step further. Steiner was an outstanding scientist, who from youth possessed a complete natural clairvoyance. He writes:

> The reality of the spiritual world was to me as completely certain as that of the physical world. But I felt a need to justify this to my thinking. I was resolved upon demonstrating to my own mind that the experience of the spiritual world had the same scientific validity as experience of the physical.

He achieved this through an intensifying of his thinking so that he could unite with the world process, drawing knowledge directly from the creative source. His concern was to refute Kant's contention that there were limits to knowledge. In achieving this merging of scientific thinking with spiritual knowledge, he stands as a prototype for our age.

In our time the 'holistic' world picture re-emerges. It is recognized that life is a great Oneness, that everything intercommunicates, and that the workings of Law within nature represent the divine intelligence manifesting in infinite diversity. The mechanistic materialism of the last century saw Man as an accident of evolution in a nature and universe wholly indifferent to him. The emerging world-picture of our time restores the sense that the Earth is indeed a living creature, an organic being, of which Man is an integral part. Here we might quote Edgar Mitchell as he came out from circling behind the moon and saw the beautiful Planet Earth again:

> In a peak experience, the presence of divinity became almost palpable, and I knew that life in the universe was not just an accident based on random processes. This knowledge came to me directly as experiential cognition. Clearly the universe had meaning and direction, an unseen dimension behind the visible creation that gives it an intelligent design and that gives life purpose.

In his book *The Medium, the Mystic and the Physicist,* Lawrence Le Shan includes an entertaining chapter with some twenty quotations, and invites the reader to guess whether they are written by mystics or physicists. The language proves to be so close that it is difficult to decide correctly.

This advanced thinking recognizes:
— That universal mind and intelligence interpenetrates everything.
— All is created by mind, including our own bodies and brains.
— Incarnation into embodiment in the gravity field of matter involves the experience of separation from the divine will.
— Self consciousness, free will and intellect develop at the price of the atrophying of the organs of perception which still in Greek days gave entry into the invisible worlds.

This is symbolized in the legend of Odysseus putting out the single eye in the brow of Polyphemus. 'Odysseus of many devices' represents the individualized intellect which can outwit the Gods. That self-conscious intellect might develop, as was the destiny of western Man, the early clairvoyance had first to be extinguished. We may find a clue as to what happens in history, by seeing that events are symptoms of a great sweep of evolving consciousness, in which the ego of Man as a spiritual entity moves steadily down and ever deeper into identification with the body and the five senses.

This is the prolonged process of the Fall. In the early fifteenth century, this descent became virtually complete and thus with the Renaissance we entered our modern era. The faculties of perception of the invisible worlds fell dormant, but a new and acute focus upon the physical world awakened. With this came a new excitement in its exploration, a release of creative energy, and an inner drive to the aggrandizement of the ego and the satisfaction of its pride and its desires. We pay for the divine gift of freewill with the experience of the loneliness of being cut off from the True Source of our being.

In Blake's words: 'If the doors of perception were cleansed. everything would appear to Man as it is — Infinite. But Man has closed himself in and peers forth from the chinks of his cavern.'

If we can acknowledge that the human soul, the spiritual entity within us, was once one with the Source, and that the long descent of Man has involved an atrophying of the organs of perception of the divine, then we recognize that, with the coming of the Age of Reason, it became virtually impossible for a man any more to find the spiritual worlds. The condition of body/mind, as it had evolved, precluded this vision. Literally the higher worlds ceased to be there any more for his looking. Therefore, very understandably, he declared that they did not exist, and that descriptions of them by earlier cultures amounted to so much superstition. The rational mind failed to see that the disappearance of the realms of spirit could be accounted for simply through the organs of subtler perception becoming dormant. Rational man therefore experienced himself in awful separation, a lonely creature who no longer belonged to nature, a chance accident in a world made up of an infinite number of separate things. The vision of wholeness was lost, and with it the sense of meaning to life.

I now take as my inspiration the important book *Man or Matter* by Ernst Lehrs, himself both scientist and outstanding spiritual researcher. He gives as his subtitle: *An Introduction to a Spiritual Understanding of Nature on the basis of Goethe's method of Training Observation and Thought.* He claims, quoting Eddington, that in order to obtain scientific cognition of the physical world, Man has felt constrained virtually to surrender the use of all his senses except the sense of sight and to limit even the act of seeing to the use (in effect) of a single colour-blind eye. Modern man's state of mind cannot of itself have any other relationship to the world than that of a spectator.

Relying solely on the analysing intellect, the masculine aspect of our thinking, and largely excluding the intuitive more feminine vision which can apprehend the living wholeness of things, we have become mere onlookers — but because of this can analyse and control nature and build up a great technology. We have now reached the evolutionary stage when we can (and must) consciously overcome the limitations of the onlooker consciousness and reassert the oneness of mind with the universal mind.

The crown of the intellectual achievement of the onlooker consciousness can be seen in the Newtonian world picture. Newton's achievement, stupendous though it is, must still be seen as the product of a stage in the evolution of intellectual thinking which precluded the blending of consciousness with the Being within the forms of nature. Hence Blake as visionary saw in Newton the great exemplar of a scientific thinking cutting itself off from the spirit. He writes

> For I with a fourfold vision see
> And a fourfold vision is given to me
> And threefold in soft Beulah's night
> And twofold always — May God us keep
> From single vision and Newton's sleep.

Newton's 'voyaging through strange realms of thought, alone' was still to Blake a kind of sleep, until consciousness could take a step which would make it possible to think with living nature and re-unite with the whole. This step was achieved by Goethe and after him carried to its consummation by Steiner in what he called 'sense-free thinking.'

Newton's thinking discovered gravity, and opened the way to the investigation of the universe on the basis of balanced gravitational pulls. But he wholly left out the possibility of a counterforce lifting towards the circumference, the exact polarity to gravity that can be called *levity*. Lehrs in his book develops the case for levity as a scientific concept, offering a bridge between intellectual science and mystical vision, between matter and the poetry of life and being. I submit this thought as a fitting opening for our conference.

Lehrs quotes John Ruskin, that great reader of Nature:

> Take the very top and centre of scientific interpretation by the greatest of its masters: Newton explained to you — or at least was supposed to explain — why an apple fell: but he never thought of explaining the exact correlative but infinitely more difficult question — how the apple got up there (Lehrs 1985, p.191).

Ruskin's remark points to a new world-conception which must be striven for today — to quote Lehrs' words:

> a conception in which levity is given its rightful place as a force polar to gravity: in which death is recognized as a secondary form of existence preceded by life: and in which, because life is bound up with levity as death is with gravity, levity is recognized as being of more ancient rank than gravity. (Lehrs 1985, p.192)

In presenting a Goethean view of the science of nature he submits that the levity-gravity relationship is to be seen as the primary polarity in the working of the laws of nature.

The conception of a force polar to gravity, actually lifting substance into refinement and metamorphosis, inevitably dropped out of the thinking of the last centuries. The time has come to reinstate it in our thinking and our world picture. Our conference itself is a symptom of this change in consciousness.

If in meditation we direct our imagination into the growing point of a plant, we may experience the emergence of form in the proliferating of new cells. Totally new matter is forming under our eyes. As this young matter moves out into leaves, the virgin shoot presses on upward into light. In a true sense we may feel that the plant or tree is being sucked up from the earth by a force lifting out towards the circumference. So equally we experience that a candle flame is operating entirely free from gravity. The downward centripetal pull cannot touch it, for it is already in the centrifugal realm of light and warmth.

The old belief in the four elements of Earth, Water, Air and Fire takes on new meaning when related to the etheric formative forces as investigated by Steiner. Air and Fire are seen to correspond with the warmth ether and the light ether, both centrifugal in their nature. They are the field of action for the levity polarity. Further condensation moves over into the realm of gravity, first into the watery element known as the 'chemical ether' and finally into solid matter or Earth, known as 'life ether.' Here we have the bridge through which creative archetypal Ideas may become operative in form in the material plane.

We watch the growth of the plant, the tender point forcing up (or being sucked up) through hard soil. Against this we may contemplate the processes of decay, the dissolution of dying plant into soil or compost, itself the matrix for new life. Do we not sense that the myriad artistic shapes of nature come, in reality, from above? Earth forces can really only break down forms into soil. This suggests an all-pervading field of creative intelligence, which contains all the archetypal *ideas*. Earth is traversed with wisdom.

As Blake wrote: 'Every rock is deluged with Deity.'

And here I must quote Steiner again:

> If we see in Thinking the capacity to comprehend more than can be known to the senses we are forced to recognize the existence of objects over and above those we experience in sense perception. Such objects are *Ideas*. In taking possession of the Idea, thinking merges itself into the World Mind at its highest potency. Such a becoming-realized of the idea is the true communion of man. Thinking has the same significance

for Ideas as the eye for Light — and the ear for Sound. It is an organ of perception.

We recognize the distinction between 'alert' matter and 'inert' matter. Where life is actively creating, as in the growing point of the plant, or in the human embryo or muscle, matter is 'alert.' Gradually the life is withdrawn from it and it becomes inert, as in wood or rock or bone. What was originally soft and supple becomes hard and rigid. Ultimately the processes of decay and dissolution set in, when the earth-forces take over and destroy form. Thus we see that the dying processes are truly deposited out of life: substance is first alive, everything is first divine Idea. *In the beginning was the Word and the Word was with God.*

Scientific intellect now establishes that all apparently solid matter is energy; spiritual knowledge shows that all energy is life and being; mystical knowledge begins to explore space by the route that Dante took, climbing the nine stages to the Empyrean, Here is the polarity to our physical exploration of space. The mystical knowledge penetrates inner space and knows that every step within is an expansion into realms of ever widening consciousness. In Blake's words again:

> The task is to open the eternal worlds
> To open the immortal eyes of Man
> Inwards into the realms of Thought
> Into eternity, ever expanding in the bosom of God
> The human Imagination.

A faculty of mind develops which can overcome the spectator consciousness, discovering that the higher self is potentially one with the whole of creation. To quote Francis Thompson:

> O world invisible we view thee
> O world intangible, we touch thee
> O world unknowable, we know thee
> Inapprehensible, we clutch thee.

But to enter these higher worlds we must achieve an inner transformation — an intensifying of thinking — lifting beyond the limitations of sense-bound thinking. Mind blends with universal mind. This is the finding of the Kingdom. This is the gateway to the wonder of the etheric plane.

We are actually experiencing this great change in our age. In us the onlooker consciousness is actively being overcome. This meeting of scientists and mystics is an aspect of the healing of that great division brought about by the coming of the Age of Reason. The centuries of separation were necessary for the fulfilling of the destiny of the West. But now a new step into the vision of wholeness is called for.

Lehrs found that, about the time of Newton, a thesis was published by the Accademia del Cimento in Florence, entitled *Contra Levitatem.* In this it was declared that a science based firmly on observation had no right to speak of levity as something of equal rank with and opposite to gravity. This was a true and logical expression of the onlooker consciousness. It is now time to re-affirm the primary polarity of levity and gravity — and the primacy of levity as the realm of light and life. It is truly the gateway to the infinite worlds of creative spirit. Perhaps therefore the slogan or theme-song of this conference should be *Pro Levitate.*

The world of the etheric, invisible to the cold intellect, can be explored by a consciousness enhanced through developed imagination and intuition. And has not Wordsworth written that: 'Poetry is the breath and finer spirit of all knowledge, It is the impassioned expression on the face of science.'

And did not Einstein declare: 'Science without religion is blind; and religion without science is lame.'

Let us do what we can to heal the wound caused by three centuries of separation; to effect this re-marriage of the masculine and feminine aspects of our thinking, and achieve this holistic understanding. This is the rising tide of the new consciousness, and behind it is the whole impulse of the angelic worlds of the Creative Source, now releasing living energy for the cleansing and redemption of our planet. This conference is a kind of invocation to the ocean of living thought and being which *fills* the universe, listening to us and speaking in our thinking if we will but be still and listen,

Our symbol will always be the Star of David. the pyramid of material embodiment, opening up to the reversed cone of light, this upper form descending to embrace and suffuse the lower. This, the Star of the Living Christ, speaks to us of the marriage of levity and gravity and pictures the true nature of Man, the human being, the HU MAN (the Egyptian word HU means the Shining One) united with Life Eternal. HU was one of the names for the Sphinx and Man is indeed the great enigma. Pope called him 'the glory, jest and riddle of the world.' The human body represents the perfect point of balance between levity and gravity as these two forces meet on the surface of the Earth. Thus this body is truly the divine temple into which the imperishable spiritual being may descend into the grave of earth.

The Greeks carved over the Mystery Temple at Delphi:

Man know thyself and thou shalt know the universe.

We recover something of the lost knowledge of Man as the measure of all things, for sacred number and proportion is built into the body. The inner sanctum of the temple, entered in meditation, is the gateway to the realms of the ageless wisdom. In our dramatic and exciting age we are called on to learn how to contact the inner Teacher that we may be guided safely through times of change, and in time become channels for the creative Source.

A supreme hope for the future of Man opens up with the prospect of carrying scientific thinking forward into the field that the seers and adepts have explored.

The Vision of Non-Duality
in World Religions

BEDE GRIFFITHS

Dom Bede Griffiths O.S.B. was born in 1906. Formerly a Benedictine monk and Prior of Farnborough Abbey, he left England in 1955 to assist in the foundation of Kurisumala Ashram, a monastery of the Syrian rite in Kerala, India. In 1968 he moved to Saccidananda Ashram in Tamil Nadu by the sacred river Cauvery (now called Shantivanam — Forest of Peace). This ashram, founded over forty years ago, was a pioneer attempt to found a Christian community incorporating traditional forms of Indian life and thought. It seeks to become a centre where people of different religions traditions can meet in an atmosphere of prayer and grow together towards the unity in Truth which is the goal of all religions. Dom Bede, whose sannyasic name is Swami Dayananda (Joy of Compassion) is the author of The Golden String, Return to the Centre, The Marriage of East and West, River of Compassion, The Cosmic Revelation *and* A New Vision of Reality. *During 1991 he travelled to Europe and the United States, carrying his vision of the need to recover the perennial world religion. The book on which he was working at the time of his death in 1993 has now been published as* Universal Wisdom. *This lecture was delivered at the 1992 Mystics and Scientists conference.*

I begin with a quotation that may be familiar to you: 'To see the world in a grain of sand and heaven in a wild flower, to hold infinity in the palm of the hand and eternity in an hour.' That is a non-dual vision of reality. I wanted to speak today of this non-dualism, starting of course where we all are in this fragmented universe, as David Bohm calls it, with everybody and everything divided. We're born into this world of division, of disintegration; this is a world of sin really, a world of disintegration, where everything is disintegrating and dividing and ending in death, and yet beyond all these divisions and fractures in our human nature that affect each person, our relations with the world, everything — beneath and beyond it all is this undivided wholeness, and it's that wholeness that we're all seeking. We're all fragmented and have this longing for this whole, for the one.

I want to suggest today how this vision of the whole has come down to us from the earliest ages; today we all need to become more and more aware of it. I'm hoping to publish a book this year called *The Universal Wisdom.* I've collected basic texts from Hindu, Buddhist, Taoist, Sikh, Muslim, Jewish and Christian sources, trying to show how this unity of vision, this converging vision of a whole, has emerged in every religion. I feel that today we have to transcend our limited religious traditions; they all have their value, but they all have their limitations, and beyond them all and pointing to it is this one reality, the whole. I want to suggest to you how we are really called today to relate ourselves to these different traditions. We want this book to be a book of meditation and not something just to read and pass on but rather something to meditate on day by day. It actually grew out of our prayer in the ashram, as for many years now we've had the custom of reading from the different scriptures. Every morning we read from the *Vedas,* the *Upanishads* or the *Bhagavad Gita,* at midday we read often from the Sikh writings, which are extremely beautiful and other sources like that; and in the evening we read the devotional poets, Hindu and Muslim and others. So I feel today we are all being called to go beyond the limits of our traditional religions and open ourselves to this universal wisdom which has come down to us from the first millennium.

The axial period of the Upanishads

In the middle of the first millennium, 500 or 600 BC, there was this breakthrough of human consciousness. There is no doubt you can almost date it — with the *Upanishads* of the sixth century, with the Buddha at much the same time, with Lao Tsu in China at the same time as Zoroaster in Persia, then Heraclitus and the movement in Greek philosophy, and of course the Hebrew prophets. It all stemmed from this period. Karl Jaspers called it the 'axial period' in human history: something happened in that millennium which transformed humanity and we all inherit that tradition. For most of us it's been closed, we've closed in on our own limited religious tradition and lost sight of the others, but today the whole thing is being opened up to us and the texts I am publishing are all in popular modern translations, so it's available to everyone today. We can all be acquainted now with this universal wisdom.

We begin with India in the sixth century — the *Upanishads* — and to me that has always been the great breakthrough. I'd remind you that the Upanishads are Sanskrit texts — 108 of them altogether — which stem from the fifth and sixth century. Of these, there are really ten or twelve Upanishads which are the classical texts for this vision of non-duality. At that moment they went beyond the sense world, the whole physical world, they went beyond the whole mental world, the world of thoughts and desires and psychological experience and they realized the absolute. It was the total experience of absolute reality, and you can't doubt it when you read these texts, they're overwhelming, and perhaps the best way to recall them is to look at the four *mahavakyas,* great sayings, that summarize the teaching of the Upanishads.

The first one is: *Pranayama Brahman,* Brahman is consciousness. That was a great breakthrough when they realized that the reality of the world is not simply just the external world of the senses, of your human experience, but that consciousness is pervading the whole universe. This consciousness they call Brahman — it's the universal spirit pervading the universe.

And then the next saying is: *Ayam atman Brahamasti.* This self, this atman is Brahman. Here they realize that when you go deep into yourself beyond your body, beyond your mind you discover this atman, this self, this spirit which is one with the eternal spirit, with Brahman. This spirit or soul *is* Brahman. It could have been a pantheistic: 'I am god,' but that's not the meaning. It's a mystical intuition that the ultimate reality in my being, my consciousness, is one with the Supreme Being, the Supreme Consciousness, and they realized that unity. And don't forget that this tradition has come down through the ages in India.

In this century there have been some wonderful mystics in our neighbourhood. There's Ramana Maharshi who was what they call a realized soul. It's a beautiful story. When he was a boy of seventeen in Madura, he was at an American school, interestingly, studying. He was a normal Brahmin boy, perfectly healthy and normal and for some reason that day he felt that he was going to die. This feeling was so strong that he surrendered himself to death, he lay down in the room where he was staying in the temple (I visited it once) and he stopped breathing, letting his body grow stiff and he said to himself: 'Now I am dead, this body is dead. Am I dead?' And at that moment he underwent a mystical death. He realized suddenly and totally: 'I am not this body, I am an eternal spirit.' And from that moment at the age of seventeen (he died in 1950 at the age of 70) he never lost his awareness: 'I am not this body, I am an eternal spirit.' He died of cancer, actually, and had an operation on his arm. When he was groaning and suffering and people were commiserating with him, he said: 'I am not suffering, this body is suffering, I am not suffering.' The eternal 'I' had formed in him. This was 1950, you see, and in our own age people have this same experience of this Atman.

Then the next *mahavakya* is well known: *Tat tvam asi,* thou art that, and it comes in the Upanishad where he takes as an example a fruit and he says: Open up a fruit and what do you see? I see these seeds. Open up a seed, what do you see? I see nothing. Then he says: In that nothing which you cannot see is the power from which the whole of that tree, the whole of the life of the world comes, that nothing, that hidden mystery at the heart and *thou art that,* in the depths of your own being, beyond all your outer form and mind and body is this hidden mystery of thou, and you are one with that, the hidden mystery of the universe.

So this is an experience which dawned in that time and it's gone on in India ever since. I think we're all being called to open our hearts to this realization. I'll come a little later to the Christian understanding. It's not very relevant in our tradition as we think of God as the Father beyond, but it's the discovery that the reality — God himself — is in me and I am in Him and *thou art that.* There is total oneness between the human being and the divine, a mystical oneness, not a pantheistic one.

Then the final saying is: *Sarva itvam Brahman asti* — 'All this world is Brahman.' You see behind again all the phenomena of this world: all that science studies, all the human reality of this world is this Brahman. And that Brahman is the reality in all beings, in all creation, in all humanity, so you always go behind all the multiplicity of phenomena to the one reality which is transcendent, beyond everything. It's immanent, it's in everything, and an interesting Isa Upanishad makes a very strong point:

> Into great darkness enter those who follow the immanent, God
> simply immanent in nature. In greater darkness enter those
> who worship the transcendent. If you separate the transcendent
> from the immanent you go into the darkness but when you
> realize the transcendent in the imminent, the imminent in the
> transcendent and the total reality you see the whole, the one,
> then you know the reality.

So that was the vision of the *Upanishads,* and I think it's our inheritance. Everyone today who is educated, who is seeking, who is on the path, should know something of this vision which dawned in the fifth century in India. I got on the track of it forty years ago or more, and I've been following it ever since, and when I came to India I realized I'd reached the source of it. It is still alive in India and now it's spreading all over the world, all over Europe and America. We're all being called to open our hearts and minds to this vision of ultimate reality, ultimate truth, the One.

Buddhism

The next stage is the Buddha at almost the same time, just a little later. And the Buddha is the opposite. The seers of the *Upanishads* realized the Atman, the Brahman, the one Reality. The Buddha started a totally negative way. He started with the phenomena, the outer world and he had this insight. All is passing, all is sorrow, all is unreal. A totally negative philosophy. All is *anikka:* it's not eternal, it's passing away. Of course science today tells us that everything is disintegrating, at every moment the cells in our body are all dissolving and recreating, the whole universe is a continual flux of energy, changing and changing. The Buddha realizes that all is passing away. As Heraclitus said: 'Everything flows, nothing is ever the same.' So this was the first great insight — the *anikka,* passing away, everything is passing away. And then as a result of that everything is *dukkha,* is sorrow, unsatisfactory, you can't find any satisfaction, it's all flowing away all the time. The moment you try to grasp it, you lose it. So we're in this world of flux and change and we mistake the flux and change for the reality.

We're always grasping, trying to hold it but it escapes us all the time, so we're all caught up in this world of sorrow.

And it's a deep insight. There's great happiness and now there's another aspect: once you realize that the world is sorrow and suffering and it ends in death, you enter great joy. The great text of the Buddhist early tradition is the *Dhammapada* and it's wonderful how, again and again it says: 'in joy we live.' Once you've realized that the world is passing away and the world is sorrow, this joy comes to you: a feeling of bliss and inner freedom. That was the experience of the Buddha. He saw through all the changing phenomena to the reality.

And the third insight was *anatta,* without an atman, without a self, without substance. It's flowing away all the time, there's no substance beneath it. This is the illusion — to think that this is a solid table and a solid body and the solid things around us. They're all flux and change, there's no substance there at all. And once he'd seen through all the phenomena, all the changing, becoming, he realized *nirvana.* And *nirvana* is negative, blowing out, like blowing out a candle, the blowing out of life. Blowing out of everything. Then when you let go of everything that is passing and changing and the sorrow of life you realize this joy. Buddha wouldn't name it, it was no god, no soul — he wanted to avoid all concepts that might imprison the mystery. He just said: 'Let go of everything and you will realize nirvana.' And of course he had the eight-fold noble path, namely the path of right living, and right thinking and right conduct, the right meditation to lead you to this experience of nirvana.

And don't forget Buddhism today is spreading all over the world. There are millions of Christians today who have discovered this great wonder in the Buddhist teaching. I remember visiting the Zen centre in San Francisco years ago, where I was told about 50% were Catholics originally and 50% were Jews. So the discovery being made beyond the personal God, is this mystery which you can't name, the nameless wonder which you'll discover when you go beyond all phenomena, all appearances, all change and becoming and you realize this wonder which the later Buddhists call *sunnyata,* the void. Buddhism is the most negative religion in the world: it's nirvana, it's the blowing out of life; it's the *sunnyata,* the void, the emptiness. Everything is contained in that emptiness: the void contains all the changing reality and the great insight of the Mahayana was that *nirvana* and *samsara* are the same. *Samsara* is the changing flux of reality, *nirvana* is the state beyond and then they realized that the one is in the many, the many are in the one — you don't lose the world, on the contrary you find yourself!

As Jesus says in the Gospel: 'He who would lose his soul shall find it.' He shall find his soul through losing it. If you cling to yourself, your soul you lose it, but you let yourself go and you find yourself, find the reality, you enter into the fullness. So that was the Buddhist vision you see, and I think we all inherit this. We all have to be Hindus and Buddhists today, we can't stand alone any more, each one grew up apart, each one thought that their religion was the only religion — most Christians think theirs is the only true religion today. Muslims think theirs is the only true religion and of course many Hindus think that Vedanta is the only true doctrine, or the Dhamma of the Buddha, but now we realize that these are inter-related and not simply separate. What we're discovering is that every religious tradition is a symbolic structure. We know reality through signs, through symbols, through words, through thoughts, through images and so on and all these are symbols or signs of the reality, they make the reality present but always under a sign. We have to go through the signs, through the symbols of all the different religions to the reality of divine mystery which is manifest in all these different traditions; and not only in the great religions, but also in the Australian aborigines, the American Indians, the African tribal religions, everywhere the one divine mystery is manifesting under these different names and forms, the images, concepts symbols, rituals, doctrines — which all have their place but they all point to something beyond. That's what we are being called to today, to go beyond the limiting symbolism of our religion and open it to the divine.

Semitic traditions

Now that brings me to the Christian tradition. Perhaps I ought to mention the Muslim tradition, but they are really all contained in this Semitic monotheism. To me this is the great problem — Semitic monotheism, whether it's Jewish or Muslim or Christian, they're all tied to this idea of one God and that God is projected beyond us. Whether he is Allah or Yahweh or the Father in heaven, we project a God, a personal God and we address him in words. We haven't faced the fact that whereas the Hindu, Buddhist and Chinese traditions were awakening to the transcendent mystery, the Semitic religions were learning (and it was an important stage in religion) to recognize one God separate from the world, separate from humanity and one alone. There's only one God, whether you call it Yahweh or Allah — so you get this monotheism which is a dualistic religion. The Semitic religion is dualistic. God is apart from everybody and everything, and He's separate from the world and the world is separate from human beings and human beings are separate from God and the world — it's all divided. That is the dualistic tradition that we've inherited. And I think today we're being called to transcend that dualism. The Indian tradition can help us to do so, but it's already present in our tradition. If you go deeper into the biblical tradition you see that beyond the separate God — don't forget the word 'holy' in Hebrew means separate. God is separate from the world, human beings are separate from God, separate from the universe, so everything is separated, dualistic.

But within the Hebrew tradition is this continual search for a unity beyond these differences and you get it first of all in the return to paradise. Paradise was the time when all was one, when the womb of the mother all was one, and then we came out of paradise. We were driven out of paradise into the divided world in which we live today. But the yearning to return to paradise, to return to the one is always there and it developed in the Hebrew tradition into the promised land — you were journeying to a promised land which was beyond this world. First of all it was Palestine. Then they realized that it's not this world at all, as the letter to the Hebrews said: we are in search of a heavenly country, not a country in this world, so we realize that the promised land is not in this world, it is a stage beyond the world, so you see the yearning to return to paradise. And then all through the prophets you get the new covenant. God makes a covenant with David with Moses and others, but these are temporary covenants. The law written in stone is something outside you and is imposed upon you, but there is a yearning for the new covenant, when the law will be written in the heart. That is what we are waiting for. The law which is love eventually is written in the heart. And so they are moving beyond the Law, beyond the written law, and beyond the law that divides. St Paul had this tremendously deep intuition, that the law is the source of sin. He was always saying that the strength of sin is in the law. You've got your natural human disposition going this way and that, the law comes along and says you mustn't do this or that, it separates you from nature, from yourself, and you're divided. You've got a moral law which you must keep and you cling to that, you have to control your nature, your passion, but Paul saw that beyond the law is the mystery of grace, where you can't control your nature, you can't manage the world as you think we can. You've got to go beyond nature and beyond the moral law and open to the transcendent mystery that is grace. You don't do it, you have to surrender and allow this mystery to enter in to take possession of you: that was the wonderful intuition of St Paul, the whole mystery of grace.

Beyond the Law

So in our Christian tradition we start with this stark moral dualism all the time, good and evil, right and wrong, heaven and hell. We're all divided, but we're all being led beyond that moral dualism into the transcendent. I think Jesus came to lead humanity out of the Law, out of Judaism into the freedom of spirit. He grew up as an Israelite under the Law. Remember at his baptism he said: 'Thus it behoves us to fulfil all righteousness.' He put himself beneath the Law, he was circumcised, went to the Temple and did all the right things. But he was taking us out of the Law, out of all this moral religion to the religion of love, the religion of grace, which I think only culminated on the cross. Jesus went beyond the Law, saying that: 'the Sabbath is made for man not man for the Sabbath.' He associated with publicans and sinners, which was a great scandal! The Israelites were a holy people, insisting that you should not have anything to do with sinners who do not keep the Law. Jesus went to have meals with these sinners; and then he associated with women, which was a very bad thing to do as well! And he even goes talking with a Samaritan woman in public when the Samaritans were separated from the Jews, heretics you should have nothing to do with. It was a terrible scandal, but he was breaking through the Law, all the time opening it up. Then I think on the cross Jesus went beyond everything. I take his words: 'My God, My God, why have you forsaken me?' I think he had to lose his God in a sense: he was still living in the Hebrew tradition and still thought of God as beyond him and above him perhaps.

On the cross he made that total surrender, he let go of God and everything, and at that moment he passed totally beyond. Now he's in that beyond, he's at one with the supreme with the beyond, which is pure love. He simply gave himself totally to love and became love itself. So this is where Christianity is leading us, beyond the law. But we still want to keep to the law — it's much safer to keep to the moral law, the rituals and all the doctrine and so on. We need laws and doctrines and sacraments and so on, but they're means, not ends, so we have to use them to go beyond them all the time, and that is our challenge today. We don't have to challenge the Church or reject Christianity. We can accept the Church and Christianity as a means to an end, as a stage in the journey which is taking us beyond. And to me it's deeply significant that Jesus always pointed to the Father, the origin, the source, through him to the Father — the Son can do nothing of himself, only what he sees the Father doing. Jesus was totally surrendered to the One, to the beyond, and on the cross he passed finally into the One, into the beyond. So that is where we are being led. I think as Christians we have all inherited to some extent the Christian tradition but we've got to go beyond its limitations. The moment we do this we begin to link up with the Hindu, the Buddhist, the Muslim and with the whole of universal wisdom.

That is where I hope we can move today; we don't have to reject church or doctrine or ritual. Wherever we stand, whatever church we belong to, we can retain what is valid in all these traditions, but we must not make them an end — that is idolatry when you make your ritual, your doctrine, your church the end, the reality. They're all symbols, signs, pointing to the reality and this concept of symbolism is very helpful. A symbol is a sign by which reality becomes present: under the signs of the sacraments the reality becomes present to us, as well as under the signs of doctrine, the trinity, the incarnation and so on. We have to recognize the limiting human conditioning and go through it to the beyond. That is what we learn from the Advaitic, the non-dualistic tradition, that beyond all dualism of the mind, images, concepts, judgements and moral laws and all the rest. This all belongs to the world of dualism and as long as we remain in this we are going to be divided. As we are now, we're going to be in conflict. However, from the moment we learn to go beyond the Law, as Paul taught us, beyond the ritual and the doctrine to the Supreme which is manifesting in the Law, then we go beyond to that point where we have this total freedom — only there can humanity become one.

We're never going to overcome these conflicts of religion —

Hindus and Muslims fight one another, Jews and Muslims fight one another, Christians, even Catholics fight Protestants. All this goes on as long as we remain on a dualistic level, the level of doctrines, concepts, judgements and so on: we will always remain divided. But once we open our hearts to the transcendent mystery to which these things lead us, we don't reject dualism, the law, the doctrine or the ritual, but these become a means, not an end, and we open ourselves to the reality itself which is calling us through all these different ways. We open ourselves to the experience of the non-dual reality. The word 'non-dual' is important. It doesn't mean it's all one — that's monism, where all differences disappear and you get a blank one. That's not the answer. The answer is not dualism, either, it's not two. But non-dualism doesn't mean two, but a mystical relationship. Our relation to the divine is a mystical relationship, you cannot express it in words, and this is our problem!. You cannot express in words what is the relation of this world, of humanity, of our own self to God, to the divine, to the eternal. It has to be experienced intuitively and that is where we are challenged to get beyond our rational lives, with all their limitations, and discover this intuitive wisdom, this universal wisdom which has come down through all the centuries, through all these religions.

Christian non-duality

I should mention very specially that we have in our Christian tradition a very strong tradition of non-dualism. It begins with Jesus himself and I always like to quote St John's Gospel. When we are studying the gospels today we have to take a critical view. The understanding today, which I think is practically universal (it's accepted in Rome, as elsewhere) is that there are three stages in the revelation of the gospel. The first thirty years Jesus is preaching the gospel in Palestine in Aramaic and nothing of that has survived, we don't know a word he said, or rather we know about half a dozen words. He raises a little girl, he says: *Talete kumi,* that is: 'Little girl arise.' That's Aramaic — just a few little phrases come through, but all the rest is lost. We don't know a word Jesus said apart from that, it was all in Aramaic. And then in thirty years from 30 to 60 AD the story and the teaching were being passed down in the churches, first of all in Aramaic, and then, as the gospels spread through the Roman Empire, in Greek. That was when the gospel was being translated from Aramaic into Greek in an oral tradition, a living tradition in the churches and then in the 60 to 90 period, when the apostles were passing away, the need for a written text came out, and that's when our gospels were written, between 60 and 90 AD. They collected up all these stories and teaching which had come down as a living tradition in the churches. There's no doubt that it links us up with Jesus and the apostles but it comes through translation and interpretation. We all recognize that you can't simply translate, you always interpret in some way. Each gospel writer interprets in his own way and they all have a very distinctive understanding. Perhaps the most important is the fourth gospel. I call it the fourth gospel because there's no evidence that it was written by St John. In fact nobody knows who wrote it; it was almost certainly written in Ephesus and Ephesus was a centre of Gnosticism.

In Gnosticism, *gnosis* means knowledge, divine knowledge, which is the same as the Hindu *jnana*, divine knowledge, which probably came from India, from the Upanishadic tradition. This divine knowledge, this knowledge of Brahman came through Persia and Egypt into Greece and Rome and was known as *gnosis*. There were Jewish Gnostics, Hellenistic Gnostics and Christian Gnostics. Most of it was considered heretical, it was not according to the strict tradition, but in the Christian church there was a Christian gnosis and the fourth gospel, the gospel of John, is a Gnostic gospel. Interestingly, it was apparently far more popular among the Gnostics of the first century than with the orthodox Christians and it was gradually absorbed into the church. So in John we have this contact of the gospel of Jesus with the *gnosis*, with the wisdom, the universal wisdom that came down originally from India. And John interprets Jesus as the *Logos,* the word, which is the word that came to the Greek philosophers like Socrates and Heraclitus, which was in Israel, and which he says enlightens everyone coming into the world.

That word, from which all creation comes, enlightens everyone coming into the world. The word of God, the divine light is present in every human being, and when we turn to him and open our hearts, the divine light begins to shine. So John interprets the whole teaching of Jesus in the light of that universal wisdom and at the culmination of St John's Gospel, Jesus prays for his disciples that they may be one 'as Thou Father in me and I in Thee that they may be one in us.' Now Jesus and the Father have this non-dual relationship, total oneness in distinction. He doesn't simply say 'I am the Father,' that would be pantheism, but 'I am in the Father, the Father in me and who sees me sees the Father.' It's a non-dual relationship, a mystical relationship, Jesus and the father, and he wants to share that relationship with his disciples in the spirit. He allows us to participate in his knowledge of the Father, that they may be one as I am Thee in Thou, that they may be perfectly one. It's the goal of non-dual oneness that Jesus is calling. And I think that is the call of humanity to enter into the divine mystery, to share in Jesus' knowledge of the Father and Jesus' love of the Father. And the Holy Spirit is that love and knowledge communicated to us — that is the goal we are all seeking.

So John's Gospel takes us right to the source, and then come the Church Fathers, all of them disciples of Plato, or neo-Platonists. They all interpret the gospels in the light of Platonism, of this divine world as Plato understood it. Clement and Origen came from Alexandria, the great school, and then we have the greatest of the Christian mystics of early times, St Gregory of Nyssa. He was a bishop, interestingly, a married man and the son of a married priest, so we had a tradition of married people in the priesthood at that time. Gregory has three stages in the spiritual life. The first is the purgative way which he compares with baptism, freeing yourself from sin, from attachment to the world to yourself. The next is the illuminative way where you open on the divine mystery, particularly the mystery of the angels of all the spiritual world — you're open to the whole spiritual world and divine revelation. In the third stage you enter into the darkness: beyond all the light is the darkness and in that darkness you experience love, you go beyond the light, the intelligence, where there is still a multiplicity, and you enter into the unity in love. In the dark. And that is marvellous intuition.

The tradition was taken up by Dionysius the Areopagite. He was a Syrian monk, they think, of the sixth century, who wrote under the name of Dionysius the disciple of Paul, and by divine providence everybody thought he was a disciple of Paul, so he practically had apostolic authority! So St Thomas Aquinas took him as the doctor of mystical theology and it went right into the Catholic tradition. Dionysius says you must get beyond every word and thought, every image and concept, beyond being itself, and enter into the divine darkness and be illumined by a ray of the divine darkness.

That Christian tradition went right through the Middle Ages to St

Thomas Aquinas and to Meister Eckhart who has been recognized as the supreme spiritual master of the Middle Ages, a disciple of Aquinas, and a leading figure in the Dominican order. Unfortunately he spoke wonderful German — he was one of the creators of the German language — and as he spoke to a popular audience and many of his sermons were taken down by disciples, we can't be sure how accurate they were. He certainly spoke very freely and — very paradoxically — he wanted to wake people up! So many of his statements were condemned and he went under a cloud, but today he's been reinstated and is seen as a perfectly orthodox master of theology and a master of spiritual wisdom, of this transcendent wisdom.

Eckhart has this profound insight that all our images and concepts of God are projections — we project an image of Jesus or the Father or whatever, and we have a concept of the trinity, the incarnation, all beautiful, meaningful symbols leading us beyond. We have to go beyond all these projections to what he calls the Godhead. God, he said, is a projection, you are the projector of God and you worship him. Beyond your projection is the Godhead, the absolute, the one. That is where we find total fullness and where we relate to all other religions.

The personal god is a tremendous problem because people project this image of a personal god — Yahweh or Allah — and they think this is the Reality, that anyone who doesn't believe in Yahweh and Allah or in the Father must be rejected, but this is your projection. It's an image, it's a name and a form. It gives a name and a form to the nameless formless One. We go through these images (they are perfectly valid, we need the personal god) but we must go through it to the Godhead, to the transcendent mystery — that is where we link up also with Islam.

The Koran is steeped in this Semitic dualism. In the literal sense, it's terrible: the condemnation of unbelievers to doom, to the fire, to punishment and to revenge. There are terrible denunciations. But the Sufi mystics in the eighth/ninth centuries developed out of the Koran — just as the Christian mystics developed out of the Bible — a wonderful universal wisdom, and here we join the universal wisdom of the Hindus, the Buddhists, the Taoists and the Christian mystics. We all meet together in that wisdom.

The great master Ibn Arabi was a master of this divine wisdom. It's a little complex and difficult to follow at times but there is tremendous depth, and he distinguished between Allah the God of belief, as he calls it, to *Al Haq,* the Reality, beyond the Godhead. Beyond the personal god is the Reality: *Al Haq.* The same happened with Judaism, they went beyond the Yahweh of the Old Testament who was so sacred you'd hardly dare to name him and came to *Ain Soph,* the Infinite, and the mystics of the Kabbalah in the fourteenth century also attained to this vision of non-dual reality.

So in Hinduism, in Buddhism, in Taoism also in Sikhism, in Islam, in Judaism, in Christianity we all converge on this transcendent mystery which has no name, no form, you can't name it, you can't conceive it: that is, the Reality beyond all names, forms and concepts. Everything we do is pointed towards the beyond and that is where we are today: we must go beyond the limiting names, forms, structures, organizations or religions and open ourselves to the transcendent mystery which is in the heart of every one of us. The divine mystery is there present in the depths of every human being and if we will let go of all the limited forms, images and concepts we discover that we are one in that transcendent mystery.

Interrelationship — the web of life

And that is where we're really being led today, and unless we get there we are going to destroy one another and destroy this planet, there is no doubt about it. If we go on as it is now with the armaments and the whole chemical industry, the planet will be destroyed and human beings will be destroyed. And there may be an Armageddon, there may be a disaster, we may not be able to stop it — but out of it is coming the new birth, the new creation, the new humanity where we are one in the One. One in the One. We don't lose our differences, we remain persons and a person is not an individual, not a separated human being — a person is a being in relationship and in the depths of our being we are all interrelated in this One. That is a non-dual relationship, or interrelationship. Fritjof Capra described the physical universe as a complicated web of interdependent relationships — the whole physical universe comes down to a complicated web of interdependent relationships, and the whole human universe is a complicated web of interpersonal relationships.

The Godhead itself is interpersonal relationship, the Trinity is the Father, the Son and the Holy Spirit, interpersonal relationship in love, communion in love that is the end of human existence, the end for each one of us. So that's where I feel we have to move today. All of us, in a particular faith or church or following, are being called to go beyond all these limitations and awake to the one Reality, and that can unite humanity today, and nothing else can.

So I think in the meeting place of mysticism and science, we're being called precisely to open up, not rejecting all that science has given us. It's very important we reject nothing that is valid in it or that philosophy has given, or theology. The key is that we open everything up to the transcendent mystery that transcends all and unites all. In discovering the Whole, we find out who we are. Who am I? I am That, Thou art That, Thou art that one transcendent mystery of grace and truth and love beyond all. Today, that is where we're all being called.

Extravagance from the Heart:
Lessons from the Mystics

MATTHEW FOX

Matthew Fox, a postmodern theologian, has been an ordained priest since 1967. He holds Masters degrees in philosophy and theology from the Aquinas Institute and a Doctorate in spirituality, summa cum laude, from the Institut Catholique de Paris. He is founder and president of the new University of Creation Spirituality in Oakland, California, and author of 21 books, including the best-selling Original Blessing: A Spirituality Named Compassion, Breakthrough: Meister Eckhart's Creation Spirituality in New Translation; The Reinvention of Work *and* Natural Grace *(with Rupert Sheldrake). He has led conferences and workshops North America, Europe and Australia, bringing his message of ecological and social justice, mysticism and blessing to eager and ever-growing audiences. In 1992 he was the recipient of the International New Thought Alliance Humanitarian Award, an honour which has previously been given to Buckminster Fuller and Mother Teresa among others. In 1995 he received the Courage of Conscience Award and in 1996 the Tikkun National Ethics Award. He published his autobiography —* Confessions: The Making of a Post-Denominational Priest, *in 1996. This lecture was delivered at the 1993 Mystics and Scientists conference.*

The issue of relating science and mysticism again in our Western tradition is, I think, at the forefront of the movement of hope in our time. This is why I was eager to respond to what this conference is about. The theme of the conference I also found irresistible, namely the 'life of the heart.' I would add that my friend Dom Bede Griffiths, who was here last year, also urged me to come.

Because ours is a time when the human race has to wake up and grow up very fast, I think it's also time to ask this question: how much of our extravagance, how much of our generosity as a species have we been hiding under a bushel? And for how long? The universe, after all, is a very extravagant place — the sun gives off so much energy every day that our entire earth runs on one billionth of the energy that the sun gives off daily, that's a lot of extravagance on the part of the sun, and that's only a near star. Are we as extravagant as the universe in which we live? The creation mystics of the Middle Ages — Hildegard of Bingen, Thomas Aquinas, Meister Eckhart, each said the following: 'The soul is not in the body, the body is in the soul.' So for my presentation tonight I want to demonstrate how the heart is not just in the body, but the body is in the heart and the mind. For example, if our minds can know billions of galaxies and search for them, then our hearts can love billions of galaxies. Everything our mind can know — if it's beautiful and good our hearts can love, but our hearts are dependent on the mind working and that's what this conference is about, science and mysticism, the mind feeding the heart and the heart feeding the mind. We are, as Thomas Aquinas said, 'capable of the universe.' That means we're extravagant, our human mind is capable of the universe.

Let me tell you a little story about science and mysticism that's quite a practical one. Last spring I was at the University of Pennsylvania in Philadelphia, and at the end of the evening there was a round table discussion. In the middle of it this fellow in his late sixties with a bald head spoke up and said 'I'm a physicist and I have a question for you,' and he laid out this very complex question on the relation of science and religion. When he had finished I said, 'You know, many scientists I've met the last fifteen years have told me they went into science because when they were eight or nine years old they had a mystical experience looking at the stars at night.' And all of a sudden his face became ten years old . He said: 'That's it, I haven't thought about that in forty-five years. That's why I became a scientist.' And he just got into his right brain so full of awe and wonder that I did not have to answer his complex question.

That's how near every scientist is to the heart, if we don't get seduced into a one way track of left-brain thinking alone. Every scientist I know is a deeply passionate person who fell in love once, usually very young, maybe with an earth worm or a clot of dirt, or a star. Eddie Kneebone is an aboriginal of Australia, and I was talking to him in January when I was there. He told me a story when we were out one night. He said: 'You know, when we look at the stars, we don't see stars, we see the camp fires of our ancestors. The stars that are near are the ancestors that left a little time ago, and the stars that are far are the ancestors from way back.' That's how close we all are connecting our cosmology to our hearts and to our histories. There's a Dutchman in our programme this year, who tells me that when he was five years old he went outdoors to look at the stars at night and instead he saw bombers coming over his city of Rotterdam. He was deprived of the heart of extravagance that every child should have and the child in every adult needs; even worse, he was deprived of it by war. By that anthropocentric failure that war always is. And the anthropocentric preoccupation that war always is, including the war going on in Bosnia and Serbia right now. The subtle victims of war are the mystic child inside the children that never gets a chance to be extravagant.

The death of the heart

One rule of thumb for dealing with a spiritual concept is to treat its opposite first — very practical advice, as we are usually more expert at the opposite. So here we have a theme — the life of the heart — what would be its opposite? The death of the heart or maybe a cold heart, a frozen heart, a dried up heart, a care-less heart, a heart without care, an uncaring heart, a passionless heart, a sentimental or titillated heart, a trivialized heart, a cold, mechanical heart. Another way to speak of the opposite of the life of the heart is to talk about or teach exclusively the life of the head. Heart has a special relationship to the right hemisphere of the body because after all it's in the body, the heart is in the body — that's a great revelation to many Westerners. It's in the left side of the body. The over extension of left brain knowledge during the modern era in the West has left us as a civilization heart impoverished, so hungry, spirit starved. And this is why the poet Rilke early in the century could tell us that the 'work of the eyes has finished, go now and do heart work.' Heart work is primary work of the 1990s for our species: we are making the trouble that exists in the world because our hearts are not living fully. We might even say that during the modern era we have committed the reductionism on the heart perhaps because heart is in the body and first we committed reductionism on the body calling it a machine (which are Descartes' exact words).

Theodore Roszak in his latest book about ecopsychology makes this point when he says:

> Thanks to the Enlightenment the body was a disenchanted object that was nothing but a machine which can be analysed into springs, levers, pistons and pumps. This is a distant origin of our conviction that the body can be improved upon by spare part surgery. The psyche came to be seen as nothing sacred because a body to which it was attached was nothing sacred.

That sentence is a bit of a shock. There is first the denigration of the body, leading to a denigration of our psyches and our sense of the sacred. The body after all does not come last in our experience of our humanity, it's among the very first experiences — a baby is busy checking those toes and fingers out, and there might be a clue here to what's missing in our own spiritual history in the West.

I have tried to reflect briefly on the negative side of this theme of the conference and I think it's enough to underscore the timelessness and importance of the theme 'the life of the heart.' Our hearts are too small, that is a most basic problem with our species today: too small to forgive. That's what's going on in Bosnia, you talk to the Serbs and ask why they are doing this — it's because it was done to them by the Muslims sixty years ago. OK, so it was. Our hearts are too small to forgive, too small to let go. Too small to undergo the grief work necessary to see more clearly so we can give birth more radically to ways of living in the world, ways of education, of worship, of politics, of economics, of relationships of all kinds, including our relationship to all the earth systems that are creative. The heart, therefore, like the mind and the body, has fallen victim to reductionism in the modern era.

Aquinas and Aristotle

I want to talk about the quadruple life of the extravagant heart. I wish to invoke the four paths of creation spirituality, using the mystics in particular because they are by their job description heart specialists.

Mystics are heart specialists. And I am specially going to call on my brother — Thomas Aquinas — because he was such a friend to scientists, such an ally to scientists. He was such an ally to scientists that he was condemned not once, not twice, but three times before he was canonized, precisely because he followed a pagan scientist named Aristotle. Aquinas insisted on bringing science right into the heart of the Christian faith, which upset the fundamentalists of his day, just as the parallel situation upsets the fundamentalists of our own day. Aquinas said that 'a mistake about Nature results in a mistake about God.' I'm not sure that there's ever been a sentence that so honoured the vocation of the scientist. Aquinas is saying, if we don't get it right, if we don't get nature right (and that's a scientist's task), we're not going to get our stories of the divine right either. Aquinas said, 'revelation comes in two volumes, Nature and the Bible.' The Bible alone is not the source of revelation, all of Nature is, all of fifteen billion years that has brought us to this moment, all the one trillion galaxies which constitute our home *oikos*, our home ecology, this is revelation. Meister Eckhart said, 'If I spent enough time with a caterpillar I'd never have to prepare a sermon because one caterpillar is so full of God.' He is standing on Aquinas' shoulders, and was deeply aware of the revelation of nature itself.

The four directions

These four paths that we will examine in the light of the extravagant heart correspond to the four directions of aboriginal prayer. The four directions of aboriginal prayer around the world are of course a way of putting our psyche into a cosmological setting. It's a way of expansion and of extravagance reminding us that prayer is not about excessive introspections, it's about extension and connecting to the powers of the cosmos. The Christian cross has this archetype to it as well, but seldom is alluded to because we are so psychologized with the spirituality of the cross in the West. The heart is not only a machine or a pump, as Vaclav Havel puts it. He has said that 'the modern age treats the heart as a pump and denies the presence of the baboon within us.' So if we mechanize the heart and make it cold, we not only destroy our extravagance as beautiful creatures but we deny our extravagance as shadow workers. We are a very extravagant species — we have it within us now to end this planet within fourteen years, as Lester Brown said, even without nuclear war. Extravagance is very much part of our psyches so we need some extravagant prayer, we need some stretching, some cosmic prayer to awaken us to our power, power which can be used either for destruction or creativity. Heart and cosmos go together, both are quasi-infinite; in fact that's what spirit means. Aquinas says: 'Spirit is our relation to the whole of the universe, our capacity for the infinite.' In leading you through the quadruple life of the heart I hope you'll be opening issues of educating the heart on the way, because that is the primary work for our species in the 1990s, to educate our ailing shrunken, shrivelled heart, or as Schumacher put it: 'The primary work is to put our inner houses in order' — to make them wet again, and green again like Hildegard of Bingen said, and red again as Native American people put it. From Native Americans I have learned that one is not a red person because your skin is red, one is a red person because your blood is red and your heart is red. We cannot pass from knowledge to wisdom without heart work, without heart education and an awakened, indeed a resurrected, life of the heart.

The full heart: the via positiva

And so the first of the quadruple paths of the heart is the full heart. The mystics also called the full heart the *via positiva*, the positive way of being filled with ecstasy and delight, with joy and gratitude. I think that in the English language the words thankful and grateful indicate there's no such thing as being half full of things. You've either been overwashed by gratitude or our hearts have not yet experienced their capacity. Thomas Aquinas from the thirteenth century put it this way: 'Spiritual joy has three levels, the first is harmony of feeling, the second is the expansion of the heart, and the third is the movement out to other things with this harmony and expansion of the heart.' The praise of God, says Aquinas, ought to come from joyfulness of the heart. — 'what we have in our heart we confess with our mouth.' So the experience of joy is not a minor thing in the spiritual life — it may in fact be the beginning of our life. Aquinas says: 'Sheer joy is God's and this demands companionship.' The creation of self began because of the joy of divinity who did not want to be alone because when you're joyful you want to share it with others, and that is why the universe exists — because of joy. That's why I took that phrase 'Sheer

Joy' as the title of my recent book on Aquinas' mysticism. Very few people have approached Aquinas as a mystic and it's somewhat understandable because he's in the scientific tradition as a scholastic. Jung says scholasticism was the first stage of Western science and there certainly was a discovery of the left brain in the Middle Ages, from Islam, which moved Western society in one direction. But what's been missed is that Aquinas was also a mystic, he brought the two together.

His greatest mystical work was written when he was twenty-six years old: it's his commentary on Pseudo-Dionysius, never translated into English — though I did so as part of my book — but it's deeply contemporary, it's all about interdependence. Aquinas says that every being in the universe has a relationship, a friendship with every other being in the universe and with God. This is profoundly cosmological work and it carries such an insight into heart that at such an early age he grounded his mysticism as a cosmological and not a psychological mysticism. This is where he breaks with St Augustine, breaks with St Bernard, breaks with that whole tradition of anthropocentrism that really began with Augustine's question: 'Am I saved?' It was not a biblical question. 'Gratitude,' says Aquinas, 'depends chiefly on the heart, the debt of gratitude flows from the debt of love.' Aquinas touches on how cosmology opens the heart up. He says: 'Isaiah says, "Spread out your tent." This means the heart must first be spread out on account of the greatness of the guest. And Jeremiah says, "Do I not feel heaven and earth? It is Yahweh who speaks".' 'Yes,' says Aquinas, 'we spread out the heart on account of the number of gifts. Isaiah says, "You will see that you will grow radiant, your heart will be throbbing and full".'

This passage gives you a feeling of what Aquinas got out of the prophets — this sense of cosmology, this fullness of the heavens, filled as they are with the divine. The whole universe is God's home. Aquinas comments on the Psalm which says, 'You should be drunk with the plenty of your house,' and adds: 'That is, the Universe.' We are here to get drunk on the plenty of God's house and that is the universe. We are here for intoxication's sake — we are here to get drunk. It's not unlike Jesus saying, 'You save your soul by losing it.' The reason we have so much addiction in Western culture — drugs, alcohol, television, shopping, money, power — is because we're not offering the authentic intoxicant which is the universe itself, God's home, which is so full of wonder and awe and mystery that it alone satisfies, it alone leads us to the house of the heart. Jesus cleared this house in driving moneylenders from the temple. Today more than ever we can realize through the new creation stories from science that the whole universe is God's temple and we have to start driving the moneylending mentality, the merchant mentality, from our hearts and minds if we are to realize what a sacred home we are in.

Another image from our ancestors refers to how Moses took his shoes off before the burning bush. We now know that it is a fact of physics that every bush is burning, not only every bush, but every being, as Professor Popp from Germany has illustrated, contains photons and lightwaves. That means we're burning, we should be taking our shoes off all day long in reverence for the sacred that's passing our way. This really came to me this Fall when I was up in Canada and I looked up a street and a maple tree was brilliant in its colours — and that's when it came to me, that a maple tree is a burning bush. It's all a burning bush. Says Aquinas, 'The freezing of the heart and the hardening of the heart are dispositions incompatible with love, the first act of love is a melting of the heart.' I love that. We can do something about our frozen and hard hearts. He says: 'A softening of the heart happens by melting, whereby the heart shows itself to be ready for the entrance of the beloved.' Faith is in the heart. The main reason Aquinas, and he's explicit about this, preferred Aristotle to Plato, is that Aristotle was grounded in the earth and grounded in nature and indeed in science, in material things, as Aquinas said, and Plato was always like some New Ager, getting off into the air and saying there's a perfect world some place. Aquinas, like the Jews, was content with the imperfection of this world, in its glory, its radiance, its Doxa, its Shekinah, which is the feminine side of God in Judaism — all this teaches that our every being is radiant but imperfect. As Adriane Rich puts it, a Jewish prophetic feminist of our day, 'we must attend imperfection school and quit chasing after Plato's ghost.' So we do find the glorious in matter itself. Augustine defined spirit as: 'Whatever is not matter.' That's dualism, cutting matter off from spirit. Aquinas said that Spirit is the *élan* — the vitality in all things, matter included. His vision of the universe was like $E = mc^2$. Energy is every place, it's all convertible, spirit is every place. So it's all revelation. As Aquinas says, 'A fool has apathy in the heart and dullness in the senses.' This is the very meaning, he says, of folly, whereas a wise person *(sapiens)* is so named because *sapor* means that we taste the quickness and savour the meats in the delicious gifts of life. Wisdom is about savouring, folly is about apathy and not caring.

The final example of the *via positiva* of the expanded heart in Aquinas should be this when he says: 'Exultation should be joy leaping out through exterior things. Joy means an interior dilation of the heart. A person who's claimed by the love of God in the beginning takes more joy later, moderate joy, but joy always expands the heart while sorrow contracts.' The expansion of the heart. That is Aquinas' agenda and I think it's our agenda for today; it's agenda for all healthy mystics. Aquinas says the Holy Spirit dissolves the hardness of hearts; this kind of work, the capacity to move from complacency and apathy to exultation is always present. Aquinas says that what interferes with inertia is zeal, but that zeal comes from an experience of beauty, an intense experience of beauty, that's where our energy gets awakened. He says we call God zealous because God knows the in-depth goodness and beauty of every being in the universe. That then is what I mean by the first step on the path to a full heart: the *via positiva*.

The emptied heart: the via negativa

The second step is what I'd call the emptied heart. The emptied heart is another path into an expanded heart and I think there are basically three ways in which our hearts are emptied. One is by meditation. Meditation gives us a silent heart, it teaches us to listen, it teaches us to be with being, not with our agendas or with someone else's agendas which might be far worse, but to be empty, to be open, to be a still and silent heart, a chaste heart, a pure heart. As Peter Deunov puts it: 'A heart as pure as a crystal.' Or as Eckhart put it: 'To return to your unborn self' — what an invitation! It's such an invitation to freedom and starting over and liberty that we can return in meditation to our unborn self. At any time. Eckhart says no matter what you've been doing all your life, all it takes is one breakthrough to re-experience your unborn self and none of your time is lost, none of our mistakes is wasted, it's all brought up into our reborn self. A beginner's heart, therefore, is a child's heart, a new heart. As Jesus says, 'Until we become like children we cannot receive the kingdom and queendom of God.' Aquinas comments on meditation as a path to the empty heart when he says:

Good people love themselves by loving their inner person.
They wish the preservation of the inward self in its integrity.
They desire good things for their inward self and they do their
best to obtain them. They take pleasure in entering into their
own hearts because they find there good thoughts in the
present, the memory of past good and the hope of future good.
All of which are sources of pleasure. Indeed they experience
no clashes of will for the whole soul tends to one thing.

That's a marvellous invitation. It's OK to be alone — solitude,
aloneness, getting to know your own goodness, your own blessing is
part of what becoming a good person is about. Aquinas says:

Wicked people do not take pleasure in their own company by
entering into their own hearts. The reason for this is that
whatever they find in their past, present and future seems evil
and horrible. They are unable to find harmony with themselves
because of the gnawings of conscience.

I talk about the alchemical task in our culture of converting loneliness
to aloneness or solitude — making something wonderful out of
loneliness and I don't think it's an easy task.

In America we pay baseball players millions of dollars a year if they
could hit three hundred, that is if they can get 30% hits each time they
get up to bat. We pay millions of dollars for 30% success. You'd flunk
out of any school with those grades. I say if in trying to convert
loneliness to aloneness and healthy solitude we batted 500 we should
be giving each other millions of dollars or at least pats on the back.
Aquinas says: 'in meditation we learn to be single hearted, we become
compact and united and more like God. This is why the heart is said
to be in the midst of the person, because God dwells in our hearts,
God is said to be in the midst, turn to your heart first of all.' It is a
foundation of your life, approach it to listen to the word of God in it.

A second way by which we experience the empty heart is by
suffering. Suffering is the experience of our heart breaking. Joanna
Macy says: 'When your heart is broken the whole universe can flow
in,' and Bede Griffiths talks about 'the yoga of despair.' He says that
many people do not experience God or transcendence in their lives
until they go through intense experiences of disillusionment, despair

and bottoming out. I thought of Buckminster Fuller just now, whose daughter died young, and he went through this great time of despair, wrestling with whether he should commit suicide or not. It was in that bottoming out that his heart opened up and he heard good news, he heard he had a purpose in life. Rilke puts it this way 'and if the earth no longer knows your name, whisper to the silent earth I'm flowing, to the flashing waters say "I am".' In our periods of suffering, darkness of the soul, bottoming out, we lose our 'I am' like Buckminster Fuller did. We lose our *I am,* our purpose for being, maybe it's the death of a relationship, a divorce, the death of a loved one. And what Rilke is advising is that the cosmos can not only absorb the loss, and the pain, but if you go there with an open heart, your own 'I am' begins gradually, silently on its own terms, in its own time, to return. If the earthly no longer knows your name, whisper to the silent earth, take it to the earth: 'I'm flowing,' just a whisper, 'I think I'm flowing,' 'I'm flowing, therefore I am.' To the flashing waters say, 'I am'; the water will give you your 'I am' back. The earth will give us our 'I am' back. All that is about the extravagant heart that is vulnerable and therefore is asking to be broken in a way. It is only the invulnerable that are not broken. Mechtild of Magdeburg in the thirteenth century said, 'Only the pure of heart suffer.'

   And still a third example of emptying the heart is the way of play, of letting go and of contemplation. Aquinas time and again defines contemplation as play, it is play because it's without a 'why,' it's without a purpose, it doesn't have an agenda; it's not about making money or rewards, it's about being and this is essentially why we are here on earth as a species. We are finding that we are a special species that can play from the time we are very tiny all the way through life, we can even play with our death, we can play with our enemies. All that is about the possibilities of awe, the possibilities of the experience of the sacred that our capacity for letting go allows to happen. The hymn in the letter to the Philippians is so powerful when it talks about how God became emptied of the divine to become a human being. It is an amazing statement in any religion, that God became emptied of the divine to become a human being. If divinity can empty itself of divinity what can humans empty themselves of? We ought to start playing these sort of games around the world, emptying ourselves of fear, emptying ourselves of resentment, emptying ourselves of our parents' hatred, the sins of our fathers that we justify our going to war for.

The fecund heart: the via creativa

The third movement of the heart that is so expansive and extravagant is the fecund heart. This is the *via creativa,* the creative way, the emptied heart is the *via negativa,* the negative way of the mystics, the fecund heart is the way of creativity. Art work is heart work. When Rilke says: 'The work of the eyes is finished, go now and do heart work, all the images imprisoned within you, for you overpowered them but even now you do not know them.' Heart work means we start paying attention to the images imprisoned within us, liberating our images, and this is why in the creation tradition, art as meditation is a primary prayer experience. Listening to images and giving birth to dance, through play, through poetry, through colour, through sound, through massage, through gardening, through ritual. Reconnecting all of our work to heart work means we pass it through our experience of imagination, our images. Circle dancing, nature, silence, rebirthing, learning to breathe, all this is about educating the heart, training the heart. Otto Rank said, 'Pessimism comes from the repression of creativity.' The *Tao Te Ching* says the Great Mother is in all of us, the Great Mother that birthed an infinite number of worlds is in every one of us, and we can do with it whatever we want — that's the announcement, the holding up of creativity, it's available to all of us. Aquinas said something absolutely parallel when he said: 'The Holy Spirit that hovered over the waters at the beginning of the world in Genesis is the same Holy Spirit that hovers over the mind and the will of the artist.' In other words we are co-creators and creation is still going on, the Holy Spirit is still working. This nonsense from fundamentalism about whether creation stopped six thousand years ago or fifteen billion years ago is absolutely silly, it's totally unbiblical. The Jewish tradition teaches that the creation is ongoing, every time a human being makes an option for life instead of death then the whole world is being renewed and made again. And this is why the early Christians coming out of the Jewish tradition could call Christ the 'new creation' because he made an option for life although it cost him his death. And the Easter memory that we are soon to celebrate next weekend is exactly about that. That the last word in life is never death, if life is well lived there is always resurrection for someone, even the supernova died but it also shared resurrection by exploding and giving off all of its elements which in turn give birth to the very elements of our bodies.

I propose that the Paschal mystery of the life, death and resurrection of Christ is not about the historical Jesus alone, but is a cosmic law. Everything lives, dies and gets resurrected if it is in harmony with the Great Mother, that is with the Holy Spirit, that is with a fecund power of the universe that is obviously still working, and working overtime.

Aquinas says 'the Holy Spirit moves your heart to work.' You see it's the same Holy Spirit that moves us to our work. It is from the abundance in our hearts that we give birth to other things, that we produce things. For prayer consists of two things, he writes, 'interior action of the heart and exterior works.' For the works to be good they must flow from the interior action of the heart. This is why Jesus says 'To they who say "Lord, Lord open the door for us," I say to you I do not know you,' because, as Aquinas said, 'they did not speak from their heart but only from their mouths.' Speaking from the heart, what a beautiful phrase about the heart chakra and the throat chakra linking up. The throat chakra is so important in our time — I listen to stories and dreams of many people and two groups especially are having amazing dreams about their throats, and that is women and homosexuals. In America, homosexual men, young men, have lost their voice for centuries, their throat chakra has been taken away. Aquinas cannot talk about the heart without talking about the throat because the throat announces what the heart has exalted in and experienced, including being torn apart. Aquinas says, 'The Psalmist says "my heart grows warm within me," meaning the heat of charity is stirred up in the heart; and Proverbs says, "Is one able to hide a fire in your breast without setting your clothes on fire?"' Aquinas says, 'Thus it is impossible that anyone hides the words of God when the heart is enflamed by love, the fire burns in my meditation.' If you want to arrive at spiritual things, says Aquinas, 'your heart must be enflamed with love of God.' The effect of excitement is that one who would prefer to be silent is moved to speak. Even introverts make themselves vulnerable when their hearts are on fire. You would prefer to be silent, but you are moved to speak. That's Moses' story — do you remember him telling God: 'Send my brother Aaron. I stutter, Aaron's a much better speechmaker'? He wanted out. Any prophet wants out. But it's like John of the Cross who wrote from his prison cell that he had nothing, 'no guide but the fire, the fire inside.' The fire moved him to take his big risk to seek escape. From frequent meditation, Aquinas says, 'the fire of charity is enkindled in our hearts.'

The compassionate heart: the via transformativa

In the fourth path in the creation tradition, the fourth side to our heart journey is that of compassion. A compassionate heart. A compassionate heart leads to compassion. And the word 'compassion' in the New Testament used very often of Jesus in Greek, means 'his bowels turned over.' That's very interesting, because it's talking about the solar plexus chakra. Previously I discussed the chakra above the heart above it, now the chakra right below the heart is the solar plexus. So if you get hit in the solar plexus with injustice, with moral outrage, the heart gets awakened and the heart gets grounded. Aquinas says, 'a trustworthy person is one who gets angry at the right people for the right reasons, expresses it in the appropriate way and for the appropriate length of time.' Anger has its place. Anger is the first stage of grief. The first stage of grief is rage, usually pictured as red, the second stage is sorrow usually pictured as black and the third is emptying transcendence usually pictured as white. We have to go through all those stages, we need rituals to lead us through all those stages today, because as a species we are in an intense grief — we are in grief over the despair of our young people all over the world, we are in grief over the condition of the earth which is so near to degradation, we are in grief over our history of racism and sexism and our history of colonialism. We need ways to pay attention.

There's a young black man who teaches in our programme who is a product of the ghettos of Cleveland. He once said to me, 'I know that white people are in grief over racism but you white people don't know what to do with it, and you're asking us black people to relieve you of that grief and we can't do it for you, we're dealing with what slavery did to us and what racism did to us.' I am talking solar plexus, I'm talking the chakra nearest the heart, I'm talking anger and guts and outrage and passion that leads to compassion.

Aquinas says, drawing on the Jewish tradition, 'Compassion is the imitation of God.' When our hearts learn compassion we are learning our divine nature and each of the three paths that precedes this fourth path opens us up to compassion, for compassion is the experience of the interdependence of all of us, all beings including God. Therefore compassion is half about celebration because when you realize your nearness to the divine in all things, the first thing you want to do is celebrate, let go in a joyful fashion. But the second half of compassion is struggle, struggle for healing, struggle for justice. Aquinas says compassion is the fulfilment of justice not its abolition. God is compassion. To be compassionate, says Aquinas, is to have a heart that suffers for the misfortune of others because we think of it as our own. In other words interdependent. Compassion is not about feeling sorry for others, dropping crumbs from the table, patting people on the head, patronizing, that's not compassion, that's altruism or liberalism, or whatever you want to call it. But compassion is realizing exactly what Aquinas said: that other people's grief is our grief. No one owns grief. The Bosnians don't own it and the Serbians don't own it. The Jews don't own it and the Palestinians don't own it. Celtic people have grief, they may not have paid attention to it, they ought to — there's no way of understanding the Jewish grief or the Bosnian grief without passing through your own broken heart. Until we pass through our hearts we are just feeling sorry for others, we're not living compassion. As Eckhart says, 'Compassion begins at home, with your own body and your own soul.' You get to compassion by sharing your stories. This is why the men's movement is so important today. Men finally paying attention to our stories of pain, suffering, keeping it in, bad relations with our fathers perhaps, whatever it is, it's holding our hearts too tight. I emphasise men because women have been smart enough to be telling their stories to each other for several decades in our generation and this is why they need men who can catch up with them spiritually.

Beyond the four directions to seven directions

Aquinas makes this amazing statement: 'The objects of the heart are truth and justice.' Truth and justice are the objects of the heart. No one can say again that Aquinas was a rationalist. His interpreters have been doing the Enlightenment, they wanted to prove how rational he was. No one can say what Aquinas says and be a rationalist. That the objects of the heart, (he doesn't say the head or the mind even) but the objects of the heart are truth and justice. These are the things to be most passionate about. Aquinas says: 'There are times in life when you must love truth more than your friends.' This has meant a lot to me in the last few years, in the last few weeks. 'We must love truth more than we love our friends.' That's passion for truth, passion for justice. Eckhart has this overwhelmingly poignant image of our work when he says: 'When you return to your origins then all your work happens from and in the heart of God.' The God heart is where we live and move and have our being and where we produce our work in the world, where we give birth to passion and compassion to truth, healing and justice. These then round out what I call a quadruple path of the heart, the four paths of creation spirituality, but in conclusion I have to add the other three directions of native American wisdom very briefly. The fifth direction is sky. Wisdom and has already been alluded to tonight, this is our movement today — from knowledge to wisdom. Wisdom includes heart, wisdom includes our creativity. As Hildegard of Bingen says, 'There is wisdom in all creative works.' And this is why the artist is so important today as an instrument in bringing forth our wisdom.

The fifth direction is wisdom depicted by eagles that fly in the sky and get a perspective on things, but the sixth direction is the ground, the roots. And here I want to say two things, one is tradition. Tradition is important. I am a deep ecumenist. I know that all the world religions at their best teach us mysticism, the Buddha nature in all things, the cosmic Christ in all, that's why we have people from all religious traditions in our faculty working together with the new cosmology teaching; but I agree with Jung when he says: 'We Westerners cannot be pirates thieving wisdom from foreign shores that it has taken, for example China thousands of years to develop as if our own culture was an error outlived.' We in the West have to demand more of our spiritual traditions because we do have great mystics like Eckhart and Aquinas and Hildegard and many others and Jesus and the prophets he comes from and wisdom literature that is so cosmological but we have been out of touch with this tradition. So that's part of grounding, learning one's tradition, as Howard Thurman, the African American mystic, says, 'The more I relate to the universe everywhere I must relate to something somewhere.' A moment like ours of awakened cosmology is not just about spacing out although that's important, it's also about grounding ourselves, otherwise we'll never return when we space out and we'll never have an influence on community or history or our social institutions unless we also remain grounded, put our roots deeper. Only then can you stretch, as far and as extravagantly as possible.

The second point I want to make about the grounding direction, the sixth direction, is sexuality and our bodies. David Lorimer alluded to our growing realization that 'the body is important to spirituality.' Let's face it — the native traditions have always known that bodies are important to spirituality, and women have always known that bodies are important to spirituality — it's men who have been shooting up into our upper chakras without paying attention to the lower ones! It's our problem in the West. That's why the drum is so important to native people, it's the first chakra, the vibration chakra bringing us into the drumbeat of the earth and the drumbeat of our mother's heart in the womb. I will show you a beautiful letter that I got a few weeks ago from one of our faculty members who does movement as meditation. He is very much in touch with this sixth direction of grounding and bodily-ness. He is taking a year's sabbatical out in the woods as a hermit this year, but he wrote this letter to me recently and I want to share it with you, it's from the diary that he's writing. He says:

> Sexual pleasure may be rewarding to one who holds in body spiritual pain. When this pleasure reaches ecstasy that is pure pleasure beyond this pain such as an orgasm, the body may be transcended, the spirit is set free. Orgasm keeps one mindful of the physical-spiritual connection and how brief is life in the body. It is in the time of orgasm that we leave the body and references to earthly time and space. It may be a time of profound spiritual healing for those who approach it with sacred intention, that is an act of beauty within the territory of the heart. We live in a culture that swings between extremes of people overworking their hearts while others are heart chaste, do not permit their hearts to fully engage in the intercourse of life. These extremes go on in places of healthy, sacred sexuality where the heart is expressing itself in concert with the genitals. We need to realize that when we restrict the body from its grateful sexual expression, its genital sexual expression, we at the same time are restricting the heart. A healthy balance between genital and heart expression keeps the body in tune throughout the spine, the axis mundi. The body as cosmos operates in wholeness with each part subject to the laws of nature that govern the body as a universe.

I think such kinds of reflections are very important in our time. It is important to recover the second chakra, our sexual chakra as an integral part connecting to the heart and also to our cosmology. The West has done a terrible job of remembering its own sacred book, the Song of Songs, the Song of Solomon in the Bible, that honours sexual love. Lovemaking on the Sabbath is a theophany, an experience of the divine. The fact that these kinds of meditation are coming out of contemporary journals is part of the healing of our time.

Finally, then, the six directions all lead to the centre and in native tradition the centre is the seventh direction: our heart. The heart of hearts where we live with passion that leads to compassion, where we truly live, this is where all life begins and where it ends. So the seven sacred directions are the directions of the quadruple heart. As Aquinas says: 'God is life, *per se* life.' So it is at the heart that we truly live and we bring all this energy of all six directions to bear on the centre of our beings.

Healing the Heart:
An Alchemy of Consciousness

ANNE BARING

Anne Baring was born in 1931 and educated in England, France and America and read History at Oxford. In 1961, after travelling widely in India and the Far East, she wrote (under the name Anne Gage) The One Work: a Journey towards the Self, *the story of her quest for the underlying unity of Hinduism, Buddhism and Christianity. For twelve years she was a dress designer with her own shop in London. She is a member of the Association of Jungian Analysts, London, the International Association of Analytical Psychologists, and the Scientific and Medical Network. In 1991, she (together with Jules Cashford) published* The Myth of the Goddess: Evolution of an Image, *an exploration of the image and mythology of the Divine Feminine from the Palaeolithic era to the present day. Her most recent books with Andrew Harvey are* The Mystic Vision *and* The Divine Feminine. *This lecture was delivered at the 1993 Mystics and Scientists conference.*

In a dreaming night scene painted by a king of Provence in the fifteenth century, the king's heart is entrusted to a knight, who will carry it with him on the alchemical quest for the treasure, the magical journey of the soul.

This talk is about the quest for an alchemy which could heal the wounded human heart, an alchemy which could transform our perception of life from lead into gold, transform it from something that is cut off from the profound depths of our soul into something that is consciously in touch with them. Such an alchemy would tune us to resonate ever more finely with the vast, humming music of the ground of being.

The heart is like an umbilical cord that connects us to the life of the whole, the greater life of the divine ground. The heart is our creative imagination, born of our instinct for relationship with that greater life. The heart generates all our quests, all our hopes and longings and will ultimately reunite us with the source from which we have come. Without the heart, without the instinct to feel, to imagine, to hope and to love, life is meaningless, sterile, dead. When we are in touch with our heart, when we are connected to our feelings, it comes alive, it vibrates, it sings.

The life-bearing energy of the heart rises like a fountain within us to nourish and irrigate the soil of soul. As with the physical heart, if the psychic heart is not in a healthy state; if one or more of its arteries is blocked, if the circulatory system is not in good order, then we cannot function at a level of optimum health. Our heart carries many wounds, and these, like blocked arteries, can restrict the flow of energy through the psychic circulatory system, leading to the impairment of psychic and physical health. Where do these wounds come from and how are they inflicted?

The evolution of consciousness

To answer this question we have to look far back into the past to understand that the kind of consciousness we now have has evolved infinitely slowly out of the ground of being. Once we were contained by it as a child within its mother. Self-awareness, the ability to reflect on our actions, to think, to reason, to focus our thought is a very recent development in relation to the thousands, even millions of years of human evolution. The potential for consciousness was buried within us, like a seed buried within the earth. Once, we lived unconsciously, instinctively, without self-awareness. The evolutionary development of the differentiation of consciousness has inflicted on us the same kind of psychic wound that a child experiences when it is born because we have come to experience ourselves as separate from the matrix that once contained us.

This long process of differentiation has been experienced by us as an exile, a fall, a state of disharmony and disunion. From it has come our present still fragile consciousness and the fears and anxieties which torment us. But the memory of fusion or union with the ground of being that we once knew, albeit unconsciously, lives on in us as a longing for reunion, for the ecstasy of belonging once again, to that greater Other. We have created all kinds of myths to assuage the loneliness and terror of separation and to re-connect us with the whole. The great sages and mystics of all cultures have tried to teach us how to dissolve the illusion of our separate existence so that we might experience ourselves in full consciousness as the divine ground, as divine being. But only a few individuals, so very few, have understood their message.

The sacred image, whether goddess or god, is so essential to us because it mediates between our present awareness of ourselves and the deepest dimension of our psychic life. It relates us through our heart to this ground. Looking back beyond the image of the father god who has been worshipped by patriarchal cultures for some two to three thousand years, we find that for some twenty thousand years before this, the mother goddess was the image of the ground. She stood for the whole invisible matrix of relationships that we call life or nature. She was Divine Life, both transcendent and immanent. She was the Divine Presence within her manifest form, continually renewing or re-generating it in a cyclical process of waxing and waning as immutable as that of the moon. The whole of life was experienced as an epiphany of her being. The cultures where she was worshipped had an innate trust in life because the Great Mother was close to them. Through her image, people were held in a state of instinctive participation with the whole. We could think of this as the first phase in the evolution of consciousness.

Then, about 2000 BC, there was a profound change in mythic imagery which suggests that about this time a new phase in the evolution of consciousness was initiated. The image of a Great Father began to replace the Great Mother. In the earliest myth — a Babylonian one — the mother goddess was murdered by the god and her corpse was split in half, one half making the heavens and the other the earth. Later, the Great Mother became the void, the deep, and in the Book of Genesis the spirit of god moved upon these waters and brought life into being. The new image of the deity reflected the idea of creative spirit bringing life into being as something separate from itself rather than emanating from itself, invisibly present within the forms emerging from itself. Creation was from the Word of the Father rather than from the Womb of the Mother and this change of imagery reflected a profound change in the way we perceived life. An older, participatory kind of consciousness that connected us with nature was replaced by one that increasingly emphasised the need to control and dominate nature. From now on the head rather than the heart becomes the focus of consciousness. With this change of emphasis in the sacred image, there is both an accelerated development of mind, or intellect, and a tremendous advance in technology and control of the environment but, at the same time, a loss of relationship with it.

The consequences of dualism

During the last four thousand years a fundamental dualism has permeated human culture, a dualism that has split spirit from nature and divided mind from body, intellect from instinct, thinking from feeling. As human consciousness evolved away from its instinctive ground, fear was constellated: fear of the instincts, fear of the unknown, fear of death, fear above all, of nature as the Great Mother who was the root of all these fears. Quite unconsciously, this fear has led to the situation where the greater part of life has been emptied of spirit: nature, soul, matter, body, instincts were gradually desacralized and the sense of life as a sacred totality from which nothing could be excluded was lost. From this time on the emphasis is on the development of power and control over nature, mythically reflected in the image of the hero killing the dragon. Human consciousness, unconsciously identified with mind and spirit, is imperceptibly elevated to the posture of a god. In our efforts to control and direct life 'from above,' and shape it to our defined goals we have assumed the mythic position of the deity we have worshipped. The idea that life is created and controlled by a power outside and above nature originates in Babylonian mythology and was transmitted to later cultures — Persian, Greek, Roman and Judaeo-Christian. We are utterly unconscious of the fact that the structure of our religion, philosophy and science rests on the foundation of a belief system which divided life into light and dark, good and evil, male and female and associated nature, matter and woman with darkness, chaos and evil.

Once, long ago, when spirit and nature were not yet sundered, we *felt* life was sacred. The Great Mother as the womb from which life emerged and to which it returned was an image that expressed the mystery of *relationship* between all aspects of life, hidden and manifest. With the loss of the image of the Great Mother, the realm of the heart, the root of the creative imagination, was increasingly dissociated from consciousness. It has had to function subliminally, unconsciously, and more and more negatively as long as nothing was done to reunite it with consciousness. No healing could take place, no fundamental shift in human values, or human understanding because everything related to the heart, to the instincts and the imagination, was devalued in relation to the rational, conscious mind. Because of the loss of the image of the Divine Feminine life gradually came to be seen as mere mechanism, without meaning or purpose. Modern science has emptied both nature and the cosmos of divinity.

I remember a dream I once had — of the Eiffel tower straddling the surface of the moon. At the time I had this dream, it referred to the sterility of my own consciousness, and the lack of relationship I had with my instinctual life but I think it may also apply to the cultural paradigm that we have today accepted. Identified with this iron construction of our control over nature, unable to trust or listen to our deepest instincts, we no longer know how to love life as divine life, how to listen to its voice, or enter into communion with its harmony within ourselves and the life around us. We no longer know how to value or create beauty, how to stand in awe of life, how to wonder. Everything has to pass the voice of the stern judge: 'Can it be proven?' before it is acceptable to us. So much of immense value to us is lost through this censorship. To sum up, because we have lost the feminine image of the Divine, we have lost touch with the instinct for relationship with life and this has deeply wounded our heart.

The Fall

There is a myth that stands at the root of Judaeo-Christian civilization that has also wounded the heart because it has ratified the split between spirit and nature as something divinely ordained. The myth of the Fall or of Exile from the Garden of Eden draws an image of human existence as essentially flawed, expelled from the divine world, blamed and punished by God for a primal sin or fault right at the beginning of our evolution. The myth was interpreted literally — as what really happened at a specific historical time, not as a myth about the birth of self-awareness and the immense evolutionary step of the differentiation of consciousness from the matrix of nature. It too has imprinted the Judaeo-Christian psyche with a fundamental dualism and a deep sense of guilt. On the one hand there is God in the Garden of Eden — the higher, divine world. On the other there are Adam and Eve exiled from the Garden and condemned to a life of toil and suffering on earth — the lower world. This image of exile and punishment has made us look for the Divine beyond rather than within life, beyond rather than within ourselves, longing to get back into the Garden but forever unable to.

How has this myth influenced our image of ourselves? It seems to me that there is a shadow aspect to our religious beliefs, an unconscious sado-masochism in the doctrine that human existence and the created order are intrinsically flawed; that suffering and death are a punishment for a primal sin and that sin has been transmitted like some fatal disease through the sexual act from generation to generation. At the root of our culture is the image of an angry and punishing Father who inflicts abandonment and suffering on his children. The negative imagery of the myth has deeply undermined our trust and delight in life, setting us against ourselves and our instincts.

With this myth as a divine model for human behaviour, how could we not blame other people in order to shift the intolerable burden of sin and guilt from our own heart on to another's? How could we protect ourselves from the belief that everyone, including children is born sinful rather than simply unconscious? Imagine the effect of this myth on generations of children and how evil was beaten and punished out of them. In the image of Eve being blamed for bringing death, sin and suffering into the world we have the root of the deep-rooted hatred and fear of women in Judaeo-Christian culture. In the image of Adam blaming Eve we are given the model for blaming and criticizing others. In the image of the angry father punishing his errant children, we have the model for a devastatingly powerful internal critic and for the persecution of others. It was to be expected that we would divest ourselves of our sense of sin and guilt by blaming and punishing others. The myth itself has given rise to great evil, compounding human suffering, splitting nature from spirit, body from mind, creating negative habits of thinking and feeling which have profoundly affected the lives of generations of human beings.

We bring into being what we believe. Who knows whether the atrocities we are witnessing in Bosnia may not have their origin in a fundamentally flawed belief system that has set us against ourselves and therefore, ultimately, against others. It might be more true to believe that we are a supremely blessed species instead of a flawed one. Perhaps leaving the Garden or the state of unconscious participation with nature was a privilege, a trust, something life urged us to undertake for its sake, with all the bewilderment and suffering it entailed. Perhaps our life, together with all life on this planet participates in that divine life we have been taught to believe is something different from us, beyond us.

The loss of soul

Jung called attention to the loss and devaluation of the soul (the unconscious), showing how the extraverted emphasis in our religious beliefs (that divinity is outside us) and the brutality of our rationalist philosophy did violence to our instincts; how our instincts were so devalued in relation to our thinking that a deep chasm has developed in the psyche between the conscious mind and the unconscious, between the head and the heart. When the realm of feeling is dissociated from consciousness and not attended to, the injured instinct becomes increasingly focused on power and the need to achieve it. When psychic injuries are not acknowledged and the flow of life is blocked, when the creative imagination is denied expression, the instincts become compulsive, ruthless, even demonic. Eventually, the waters of the soul dammed up behind the fragile wall of consciousness burst through it, destroying everything in their path. It is not because human beings are intrinsically evil that they create evil, but that they cannot survive the dissociation at the core of their being. Everything we call evil, the huge aggression in today's world comes from the wounded heart of the culture and the unrecognized psychic suffering of millions of individuals.

Jung saw that our consciousness has become one-eyed, a monolithic consciousness that sees only the surface of life and takes that for the only reality. He saw that the neglect of our inner life, our deep instinctual needs and wisdom, had led to the situation where, as in the Grail legend, the territory of the soul is in the grip of a terrible drought. No-one understands any more what the landscape and the language of the soul is like; no one can read the images. They are like hieroglyphs whose key has been lost. He saw that there was an immense work to be done in recovering the soul and that time was very short.

He saw that the decay of the old god-image and the disintegration and barbarism of this century heralded a new phase in the evolution of consciousness, one which would recover the feminine value and the dimension of the soul. This transitional phase of the loss of the old god-image is extremely dangerous. In our fear and confusion, we may lose the priceless attainment of the level of consciousness and insight we have struggled for millennia to reach. Because of the polarization of rational mind and instinct within us, everything around us is polarized: on the one hand there is the anarchic impulse to overthrow all kinds of authority; on the other, repressive authoritarian tendencies, fundamentalism of all kinds which attempt to resist any impulse for change. In its fear and confusion, and in its archaic habits of response to danger, the instinct may turn demonic, threatening the delicate fabric of civilization with disintegration and the descent into barbarism. Religions that did so much to create civilization, are powerless to prevent this regression, perhaps because they are unable to recognize their authoritarian shadow and because they cling to an image of the divine which projects divinity beyond human existence. Belief will not sustain us through this perilous phase. We need insight into why this situation has arisen, knowledge of our nature, understanding of our instinctive habits of behaviour, above all the integration of thinking and feeling.

Pathologies of consciousness

The image of the minotaur which Picasso drew in 1935 shows the destructive power of the instinct. The fragile little girl holding the bouquet of flowers and the two girls above stand perhaps for the heart — mute witness to the terror and the brutality of our times. But the child holds up the light that reveals the situation. Does she stand perhaps for the child in all of us who sees what we cannot see. If we could truly see the wound in our heart and feel compassion for it, there might be a hope of change. But in our fear we may continue to react blindly and unconsciously instead of moving forward on our evolutionary path.

This has been the century of concentration camps, hydrogen bombs, human sacrifice on an apocalyptic scale and of theologies of power which aim to manipulate and control both mind and body. Where does the pathology of our consciousness reveal itself? How is the creative imagination perverted and used to destroy life? Prisoners of archaic tribal responses, we spend a trillion dollars a year on arms. Prisoners of fear, we justify the invention of weapons of destruction that can exterminate millions of human beings by remote control. If we knew that life is one and indivisible, would it be so easy to destroy our enemies since, in essence, they are ourselves? If religion had taught us that the body was sacred and had not itself tortured and murdered in God's name, would it be so easy for us to atomize it? We seem, in a state of unconscious identification with God to have assumed the power of the godhead itself, using the elements of life to destroy life. This malignant aggression (Erich Fromm's phrase) begins at the point where the heart is paralysed, frozen, where our psychic life cannot grow and begins to wither and die. At this point the attraction to sadistic behaviour and death begins and with it the addiction to the intoxicating power to inflict suffering and death on others.

Healing the culture

How could this pathology be healed? Once, long ago, we felt contained within the ground of being in the image of the Great Mother. Then we learned to fear and obey the image of the Great Father and to sacrifice the flesh to the spirit. Now we are asked to restore the feminine dimension of life to its former sacrality, so we may heal the dissociation in the soul. This means loving and honouring and valuing human existence on this planet, relinquishing the negative imagery of the myth of the Fall, becoming aware of all the rigid patterns of belief which block the flow of life in the individual and in society. It means restoring matter and the physical body to the realm of spirit, understanding that they are sacred because they are the manifestation of divine life. It means learning to cherish them instead of trying to manipulate and control them, learning to cherish ourselves as an infinitely precious vehicle of life. This focus on healing our inner life is not introspective selfishness as we have been taught for so long. It is loving and serving the divine life that we are.

Only in this century has our attention been directed towards the suffering hidden beneath the surface of consciousness. The new insights offered by psychology are a great advance or increase of consciousness for they offer us the opportunity of freeing ourselves from unconscious habits of behaviour, personal habits as well as tribal habits. Now, as the old paradigm fractures and dissolves, a new conception of reality is struggling to be born. A new holistic approach to the Earth and to the body as the manifest aspect of the soul reflects this new vision. New discoveries in science and medicine, new and experimental forms of healing are part of this impulse. Where religion has judged and condemned negative patterns of behaviour as sinful, our growing insight into our nature can understand them as symptoms of psychic injuries, of deep injuries to the heart. This new impulse is grounded in compassion for life, compassion for ourselves as participants in a divine drama and the realization that the healing of the culture begins within ourselves. It is helping us to see that people are not bad but sad and that sadness can lead to badness and even to madness and to all the patterns of evil that are so resistant to our efforts to eradicate them.

This descent into the realm of the instinct is not without danger. Psychology that offers itself as a redeemer can easily degenerate into authoritarianism or facile technology. As we penetrate deeper and deeper below the surface of consciousness into the neglected dimension of the soul we reach the molten lava of long-buried emotions, the hidden turmoil of individual and collective suffering. Healing comes with the recognition, acceptance and transformation of powerful and frightening feelings. This is the direction that could lead us to the further evolution of consciousness. Where religion has emphasised repression and sacrifice, conformity to collective beliefs, guilt and punishment, the new approach does not judge or condemn but seeks to listen and to heal. Instinct is the tumultuous energy of life. It can never be controlled by consciousness; but it can be transformed through insight and compassion. If you tell someone he or she is bad, sinful, you will not heal the heart. You will compound the pain and rage in the unconscious. Compassionate insight can offer release for intolerable pain buried beneath years, even centuries, of repression. Slowly, it can free us from bondage to the collective habits of belief and behaviour that have blocked our true response to life. With infinite care, it can dissolve the false self — the defense we construct against suffering and fear and regenerate the true self, the quintessential treasure of the heart.

Creating a relationship between consciousness and the deeper instinctive dimension of the soul works an alchemy within us that is best expressed in this image of the vision in the forest. Here, the goal-oriented consciousness we know and live by encounters the mysterious dimension of the soul. This encounter brings into being a different relationship to life, a different attitude to it, a different way of living which one might call the Way of the Heart. It takes many years and infinite patience and trust to accomplish. It is about becoming aware of oneself as innately divine instead of innately flawed. It is about becoming the humble servant of life, devoted to caring for it, healing it and freeing it from our archaic fear and violence towards it. It is about retuning oneself so that one begins to resonate with life, harmoniously, ecstatically. All this is revelation.

The wounded child

This brings me to the last part of this talk and to the image of the child, for the child is both our past and our future and it is only through our understanding of the child's needs that we can hope to change the present. What wounds the heart of a child? There is a general belief that children are resilient, tough, able to survive the most atrocious experiences. But my experience with my patients suggests that this is not true. The child may survive physically and intellectually, may be able to hold its own in the world, but the wound to the heart will show in its close relationships; in the way, as an adult, it treats its partner or its children, and in depressions, obsessions and compulsive behaviour of all kinds. It will develop a defensive carapace, a false self, in order to survive the pain of its experience and may believe this false self is its true individuality. The false self in league with the superficial goals of the culture, will drive the person to seek greater and greater power for to be at the top or in control of other people is to be beyond the reach of the child's sense of powerlessness and worthlessness.

The psyche of the child is like warm wax. Its sense of self is barely formed by the time it reaches adolescence. It is impressionable, fragile, sensitive, vulnerable. What it absorbs from the atmosphere of the home and the wider environment of school and society, is imprinted indelibly on the memory. Children without a stable and happy home, children who have to survive in a brutal or depraved environment, often witnessing the emotional or physical cruelty inflicted on one parent by another, children who are exposed to the anger, lust, cruelty or the rigid belief system of their parents or step-parents, are like a baby thrown into an abattoir. They have little hope of psychic survival. Indeed, as someone has written, they are the victims of soul murder. As they grow up, the memories of intolerable pain are repressed into the deeper levels of the unconscious, into the muscles and nervous system of the body where they may manifest as illness. There may be no recall of the actual circumstances or traumas that wounded them. Later these repressed memories are re-enacted in destructive or self-destructive scenarios which are a kind of code language telling the story of what happened to them thirty or forty years earlier.

Now the tragic fact is that in every case, the child blames itself for what has happened in the same way that humanity in the myth of the Fall blamed itself for the experience of suffering and death — believing it must have angered the heavenly father to have been so punished. Children depend absolutely on their parents for their survival. A child has no reflective consciousness. In a situation of terror or punishment or abandonment or if it has to witness the cruelty of one parent towards another, it feels unbearable fear and pain. But it can neither flee nor fight, so the instinctive response to danger is suppressed. To explain the situation to itself, it says, 'I must be bad for this thing to be happening to me.' It takes the guilt upon itself. The deep conviction that in some way one is responsible for any of these disasters and therefore guilty, establishes itself in the soul as an internalized inner critic, a persistent negative voice — even a demonic voice — which undermines and destroys one's feeling of value and may destroy one's life. This deeply unconscious internal negative voice is the root cause of depression and other compulsive patterns of behaviour such as alcoholism, anorexia and bulimia, drug-taking and promiscuity, and the violence, cruelty and depravity that are increasingly seen in our society. All these are symptoms of original pain, not original sin. All are patterns of self-destruction. Because soul and body are essentially one matrix of energy, the unacknowledged suffering of the heart may place great stress on the immune system of the child, preparing the way for illness and disease later in life. Yet, as many people are discovering, the breakdown of the body may be the beginning of healing the heart of the child they once were.

When the child enters the wider world of society already damaged by the home situation, and finds an impersonal, frightening environment and a curriculum devoted only to achievement, where there is no beauty or poetry or mystery, no welcome for the heart, it will again be traumatized and the imagination will be distorted into negative fantasies. The pathology of destructive violence presented on television, film and video increases the sense of fear and powerlessness. What children watch, night after night is the spectacle of the desecration of the soul. Consciously, they may say they don't copy the negative mythology they see, but unconsciously the instinct sees violent, threatening images and it identifies with the aggressor as the only way to survive.

Children whose feelings did not matter to their parents may, as adults, ignore their own feelings and those of others. Compulsively, in addictive or manipulative behaviour of all kinds, they may repeat or re-enact the original trauma by attracting to themselves situations or relationships that punish them, traumatize them. They may also, by unconsciously identifying with the aggressor who wounded them, wound their own chosen victims — always someone weaker than themselves, making *them* suffer the intensity of the pain they once had to endure themselves.

The work of healing and transforming the wounded heart is what in alchemy is called 'drawing the dark matter out of the sea.' It is creating a relationship with the shadow side of ourselves, the part that is hidden from our rational consciousness, the part we are frightened or ashamed of acknowledging because it has been named as evil and sinful and because it has the power to control us. If this work is not done, if we fail to understand and heal the wounds that lie behind the violence in society, there is a real danger that the instinct will indeed overwhelm us as it is doing at this moment in Bosnia. The more dissociated this dark side of ourselves becomes, the more dangerous and uncontrollable it is when it breaks through as some violent form of behaviour. The wounds of the past, the negative beliefs and actions of the past are reincarnated in the ferocious hatreds and cruelties of the present. The life of the planet and the continuation of our own life on it depends upon the growth of our consciousness and our ability to understand and redeem the dark aspect of our own nature.

Healing the heart

How can we heal our heart? The child is our conduit to the heart. We can seek out this child in ourselves who was abandoned, rejected, terrorized, tortured and paralysed with fear or left bereft by a catastrophe which broke its heart. The child is the key to healing all our habits of aggression. If we can heal the child in ourselves, if we can melt the long-frozen capacity to feel, if we can transform fear, rage and guilt into trust and delight, the heart will begin to heal. The soil of soul, so long parched and dry, is watered by the flow of released feelings. Regenerated, it becomes rich and fertile. The imagination begins to function creatively instead of destructively. Ideas appear, take root, grow and come to flower in creative work that nourishes both oneself and the community, so returning transformed pain as compost to the psychic earth of humanity. Life then becomes the companion who is a constant help and support and guide. Like Tobias with the Archangel Raphael, we can discover how to trust and communicate with life as with a guiding and directing presence, above all, as a friend.   The child who is the artist, the poet, the musician and mystic at the heart of each one of us, the child who is the true creative nucleus of the individual, who is our vital connection to the ground of being, begins to feel, begins to come to life, begins to trust life, no longer fearing catastrophe, begins to feel happy. Then a miracle takes place. The person imprinted with guilt, whose internal voice said 'I hate myself' and whose actions said 'I hate life; I hate other people' begins to say 'I love myself, I love life, I love other people.' The love flowing from the healed wound in the instinctual life, the life of the heart, grows and spreads and expands. And so it happens that the lead of a crippled and tortured heart is transformed into the gold of a loving and compassionate one. This is alchemy — the recovery of the heart's capacity to love and the most precious treasure of the soul.

Resurrection, as shown in the marvellous painting by Piero della Francesca, is about a slow transfiguration of consciousness, a gradual experience of revelation offered by a deeply compassionate relationship with life, lived at all levels of our being. Healing brings transfiguration. Healing is discovering how to live life in a different way, in trust rather than fear, learning how to relate to life as partner and lover. It is about falling in love with life, about 'following our bliss' — Joseph Campbell's phrase which describes the rapture of being alive, following the longing impulses of the heart in whatever direction they lead because these alone guide us to realize life's intention for us. In following the impulsion of the joyous heart, we are doing life's will. The happiness of the heart is released when the guilt, anger, envy and self-hatred originating in childhood suffering are redeemed. The feeling of happiness grows as one begins to experience the revelation of what life is, and begins working consciously and deeply with it, as a contribution towards restoring the whole damaged fabric of life both within us and without.

Looking at the image of the Coronation of the Virgin, healing the heart to me is about raising everything to do with the rejected feminine archetype to consciousness, resacralizing it, crowning it as Christ is crowning the Virgin with our insight and understanding. It is about learning to love and understand ourselves and therefore others at the deepest level. It is about transforming our darkness instead of projecting what we fear and reject in ourselves on to someone else. It is about cherishing in every sense: cherishing the heart, cherishing and healing the once and future child within us, cherishing the time given to us as our life in order to discover our true direction and who we truly are; cherishing the body which has been sacrificed for so long to our distorted image of spirituality; cherishing the lives of the people who have been given into our care; cherishing the planetary life which is the great field of all our endeavours.

A culture grounded on extraversion alone will not survive because its values are too shallow to sustain it through the kind of crisis we now face. But there is a new consciousness coming into being — I will call it quantum consciousness — prepared and mediated by many thousands of individuals. With it, all things are possible. With its help, we can change our image of reality and the crystallized habits in which we are imprisoned because we don't know how to trust the heart and the imagination. The answers to our questions cannot come from the incomplete consciousness of the intellect but from a deeper revelation that may be born from our instincts, a new mythology of the whole of life as a divine unity. There is, in this new myth, no essential distinction between transcendent and immanent life; as the mystics have always told us, the distinction and the dualism are in our distorted perception of reality. The divine is what we are. We are eternally in the divine. This revelation above all others may heal our heart.

Editor's Note:

The Mystics and Scientists Conferences

It was not an easy task to make the selection of lectures for this book. In twenty years of conferences there have been over one hundred lecturers and nearly two hundred lectures. On looking through the titles, five main areas emerged: cosmology, physics, biology and the Gaia hypothesis, consciousness and psychology, mysticism and spirituality. The titles of the conferences themselves reflect these themes. The first one was simply called 'Mystics and Scientists' and was billed as 'The May Lectures,' which themselves had been founded in 1974 and taken over by the Scientific and Medical Network in 1977. The May Lectures continued their existence under Network auspices as a separate entity from 1979 onwards and have now metamorphosed into the 'May Dialogue.' Malcolm Lazarus and Sir George Trevelyan of the Wrekin Trust masterminded the initial conference. The first programme makes the following point: 'The scientific mind can pass through the door of the infinitesimal into the infinite. The mystic by going within and entering a state of sense-free perception can discover that he is a part of all that is. The sub-atomic physicist using methods of observation that go beyond ordinary perception, experiences reality as a harmonious organic whole.' It goes on: 'the importance of this scientific breakthrough into a model of reality that is essentially spiritual in nature cannot be overstressed.' This general theme was carried over into the second conference and reflected in the titles given by Lawrence Le Shan ('The Medium, the Mystic and the Physicist: Three Roads to the Same Reality?') and by Ravi Ravindra ('Science as a Spiritual Path.')

The next five conferences explored various aspects of consciousness and its evolution. The past few years have seen a veritable explosion of interest in consciousness studies. International conferences are constantly advertised and the *Journal of Consciousness Studies* is well established. It is striking to see such titles as 'The Science of Consciousness' and 'Frontiers of Consciousness' as early as 1980–81. As the programme of Mystics & Scientists 3 put it: 'It is here at the junction point of the universe of outer space and the universe of the inner world that science and mysticism are meeting.' The eighth and ninth conferences brought in music and mathematics, drawing on our Pythagorean heritage and relating it to architecture and rhythms within the universe. The tenth conference came back to the original question of unification and the search for wholeness. The following two conferences focus on biology, the Gaia hypothesis and their implications for the evolution of consciousness.

At this point the conferences became my responsibility, and were subsequently taken over by the Scientific and Medical Network in 1992. The first two themes addressed the nature of the self and of transformation, especially in the light of the emergence of chaos theory in the late 1980s. The 1992 conference on the nature of light was a high point in the series with the extraordinary contribution from Dom Bede Griffiths in the year before his death. It was attended by well over four hundred people, so that a large overflow hall became necessary. In 1993 the theme was 'The Life of the Heart' with Dr Larry Dossey and Fr Matthew Fox. 1994 saw a return to the theme of transformation with consideration of the various meanings and levels of birth and rebirth. In 1995 we took one of the five key elements — water — and considered recent scientific and medical research as well as its symbolic meaning. The 1996 theme was the controversial one of the nature of energy. This term is used in a precise way by physicists, who resent its apparently more metaphorical use by healers; healers in turn respond that no other word is adequate, and can point out that physicists hijacked the term which originally had a wider meaning. 1997 saw the twentieth anniversary so it seemed appropriate to return to the main theme of the conferences by taking the title 'The Spirit of Science and the Science of the Spirit.' The theme in 1998 is 'The Breath of Life.'

Mystics and Scientists conferences are organized every April by the Scientific and Medical Network, an informal international group that aims to deepen understanding in science and medicine by fostering both rational analysis and intuitive insights. It has over 1,900 members in more than fifty countries. It questions the assumptions of contemporary scientific and medical thinking, so often limited by exclusively materialistic reasoning.

A list of tapes and videos of conference lectures is available on request. Details of membership and conferences can be obtained from our website or from:

The Director
Scientific & Medical Network
Gibliston Mill, Colinsburgh, Leven, Fife KY9 1JS
Scotland, UK.
Tel: +44-(0)1333-340492. Fax: +44-(0)1333-340491.
Email: SciMedNetwork@compuserve.com
Website: http://www.cis.plym.ac.uk/SciMedNet/home.htm

Index